General Chemistry Laboratories

A Freshman Workbook

Second Edition

Simon Bott
Russell Geanangel
University of Houston, Department of Chemistry

ISBN 13: 978-0-7575-3145-3

Printed in the United States of America
16 15

CONTENTS

Acknowledgments v

1. Introduction to Measuring Techniques 1
2. Properties of Matter 9
3. An Introduction to Chemical Reactions 15
4. Copper Chemistry—A Series of Reactions 25
5. Preparation and Properties of Gases 31
6. Gas Laws 41
7. Determination of Molar Mass by the Dumas Method 53
8. Thermochemistry 61
9. Light from Atoms 71
10. Periodicity and the Periodic Table 79
11. Properties Associated with Changes in Physical State 87
12. Determination of Molar Mass by Freezing Point Depression 93
13. Organic Chemistry 101
14. Kinetic Properties of a Chemical Reaction 107
15. Determination of an Equilibrium Constant 119
16. Introduction to Acids and Bases 129
17. Acid-Base Titration Curves Using a pH Meter 137
18. Slightly Soluble Salts 147
19. Electrochemistry I. Galvanic Cells 157
20. Electrochemistry II. Electrolysis 167
21. Synthesis of Aspirin 175
22. Separation by Paper Chromatography 181
23. Redox Reactions 187
24. Acid-Base Titrations: The Analysis of Vinegar 193
25. Antacid Titration 199

ACKNOWLEDGMENTS

We would like to thank the following people for their help in preparing this laboratory manual:

William Abbott, Cathy May, Samara Kerawala, Alian Garay, Hung Nguyen, Ada Kyriasoglar, Orly Klein, Victoria Claus, Rosanette Luther, Sipra Gohel, Ariane Arnold, Tanya Brown, Edward Galvez, Suboohz Hasan, Chris Tyler, John Venezia, Weihan Kan, Richard Perry, Sujata Patil, and Caset Robertson for their help in proofreading. Any residual mistakes are our fault!

Students at the Universities of Houston and North Texas who have worked with previous laboratory manuals and tried to show us how not to run a laboratory course!

Our families and colleagues who have tolerated us during the time of writing!

Russ Geanangel
Angela McGuffey
Houston, 2006

Highest Possible Score:
Lab (prelab + data sheet + post lab): 75
Quiz : 50

Introduction to Measuring Techniques

Introduction

The purpose of this experiment is to introduce you to a variety of methods of measuring amounts of solids and liquids.

Chemistry is an empirical (experimental) quantitative science. In other words, most of the experiments you will do involve measurement. Over the semester, you will measure many different types of quantities—pressure, temperature, pH, *etc.*—but the most common will be the amount of a substance.

The substances you will encounter most often will be in either solid or liquid form. It is, therefore, essential that you know how to determine how much of a solid or liquid you have. The amount of solid is measured by mass and the amount of a liquid by mass or, usually, volume. Today, you will learn how to use the most common lab equipment for mass and volume measurement.

Procedure

A. *Identification of Apparatus*

Find the following pieces of equipment in your drawer or around the lab and sketch them on the data sheet:

(a) beaker
(b) flask
(c) watchglass
(d) Bunsen burner
(e) pipet and pump
(f) buret
(g) graduated (measuring) cylinder
(h) test tube
(i) scoopula

B. *Balance Use*

In these general chemistry laboratories, we only use easy-to-read electronic balances—saving you a lot of time and the teaching assistants (TA's) a lot of headaches. It is still important that you become adept at the use of these, however.

Three aspects of a balance are important:

(a) The on/off switch. This is either on the front of the machine or on the back. Its use is obvious.

(b) The "Zero" or "Tare" button. This resets the reading to zero.

(c) CLEANLINESS. Before and after using a balance, ensure that the whole thing is spotless. Dirt on the weighing pan can mess up your measurements, and chemicals inside the machine can damage it.

1. Turn the balance on.
2. After the display reads zero, place a piece of weighing paper on the weighing pan.
3. Read and record the mass (*1—the number in italics refers to the line number on the data sheet*).
4. Add about 1/3 of a teaspoonful of the solid into the pan.
5. Record the mass (*2*).
6. To determine how much solid you have, simply subtract the mass of the weighing paper (*1*) from the mass of the paper and solid (*2*). Record this mass in line (*3*).

You have just determined the mass of an "unknown amount of solid." A second, slightly more complete use of the balance involves weighing out a pre-determined amount of a solid.

7. Repeat steps 1–3 above, using a new piece of weighing paper (*4*).
8. Press the Zero or Tare button. The machine should return to zero.
9. Spoon out about 1 g of the solid (*5*).

The use of the zero/tare button is obvious—it eliminates the need for subtraction.

C. Dispensing Liquids

When working with liquids, as discussed above, we usually describe the quantity of the liquid in terms of volume [usual units being milliliters (mL) or cubic centimeters (cm^3)]. We use three types of glassware to determine volume—buret, pipet, or graduated cylinder.

Before discussing these individually, there is one thing common to all.

10. Examine each piece of equipment. Note that the sides of each are graduated. You can read each with the precision of half of the smallest division.
11. Put some water into the graduated cylinder. Bend down and examine the side of the water level. Note that it has a "curved shape." This is due to the water clinging to the glass sides and is called the meniscus. When reading any liquid level, use the center of the meniscus as your reference point.

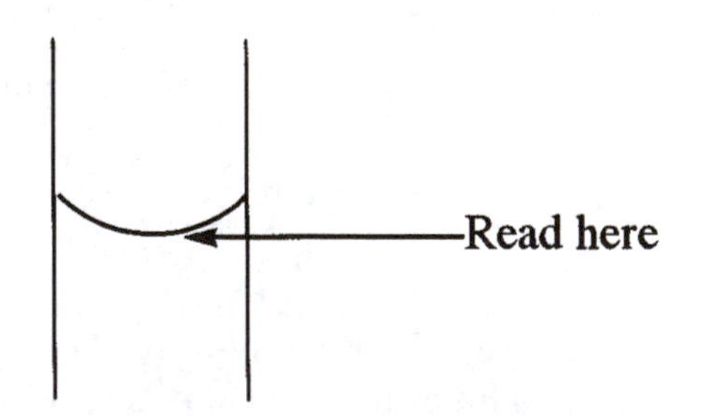

a. Graduated Cylinder

12. Look at the graduations on the side of the cylinder. Note that they go from 0 at the bottom and increase upwards. Thus, to get the mass of 6 mL of a liquid from a graduated cylinder:
13. Add water up to the 6 mL line as accurately as possible.
14. Dry your beaker and weigh it (*6*).
15. Pour the 6 mL of water from the cylinder into the beaker. Reweigh (*7*).

b. Pipet

16. Look at the graduations on the pipet. You may find either that 0 is at the spout end or at the top of the pipet. You should be aware of how these graduations go when using each pipet.
17. Half-fill a beaker with water.
18. Depress the plunger on the pipet pump and then attach the pump to the top of the pipet. Put the spout of the pipet under water and turn the pump wheel clockwise. It should draw the water into the pipet.
19. When the water level is past the last graduation, remove the spout of the pipet from the water.
20. You can dispense the liquid either by turning the wheel counter-clockwise (slow dispensing) or by pressing the lever on the side of the pump (fast dispensing). Always run some liquid into a beaker or flask that you are using as a waste container in order to leave the level at an easy-to-read mark.
21. Add 6 mL of water to a preweighed (*8*), dry beaker and weigh (*9*).

c. Buret

22. Attach a buret clamp to a stand, and clamp a buret.
23. Examine the graduations. Note that 0 is at the top.
24. Using a funnel, add about 6 mL of water. To do this, first lower the buret so that the top is easy to reach.
25. Run a little water from the buret into a waste container. Then remove the buret, turn it upside down and allow the rest of the water to run into the container (you will have to open the tap to equalize the pressure).
26. You have just "rinsed your buret". This should be done every time before using a buret—first rinse with water, then repeat the process using whatever liquid is needed in the experiment.
27. Fill the buret to any convenient level. It is a good technique to "over fill" and then allow liquid to run into a waste container until you reach the appropriate level.
28. Dry a beaker and weigh (*10*).
29. Measure 6 mL of water from the buret into the beaker and reweigh (*11*).

Properties of Matter

INTRODUCTION

Matter is anything that has mass and occupies space—*i. e.*, everything you see and a lot of things that you don't see. In order to distinguish between different types of matter, we consider their properties or characteristics.

Properties are of two types—intensive and extensive. Intensive properties are those that do not depend on how much of a substance is present (for example, whether the substance is a solid, liquid or gas under a given set of conditions), while extensive properties do depend on the quantity of a substance (such as mass or volume).

Properties can be further sub-divided into physical and chemical. Physical properties are those that may be investigated without changing the composition of the substance, whereas chemical properties describe how the substance may change composition—either spontaneously, or in combination with other substances. Thus, boiling alcohol to give gaseous alcohol is a physical change, whereas burning alcohol with oxygen to give carbon dioxide and water is a chemical change.

In today's laboratory, you will examine various properties of two liquids. The objectives of the experiment are:

(1) to observe how different substances have different properties;

(2) to understand the different types of properties;

(3) to familiarize yourself further with the lab environment.

PROCEDURE

A. Physical Properties

In this part of the experiment, you will examine a variety of physical properties of two liquids. By comparing your observations to a list of properties of "known" substances, you will be able to identify your particular samples.

(i) Appearance and Odor

1. Obtain about 20 mL of each of unknowns A and B in different 100 mL beakers. Describe their appearance on the data sheet (*1*). Use this 20 mL of each for the rest of the experiment.
2. Raise the beaker containing A until it is level with, and about three inches away from, your nose. Wave your hand over the top of the beaker to waft the vapor toward you. Describe any odor (*2*).
3. Repeat for the beaker containing B.

(ii) Density

4. Take a clean 150 mL beaker, place a test tube inside it and weigh the combined apparatus (*3*). Pipet about 2 mL of unknown A [record exact volume on data sheet (*4*)] into the test tube and reweigh (*5*).
5. Determine the mass of liquid (*6*).
6. Calculate the density in g/mL (*7*)—see sample calculations if unsure.
7. Repeat the above procedure once more for unknown A and twice for B.
8. Calculate the average density of each unknown (*8*).

(iii) Boiling Point

9. Assemble the heating arrangement shown in the diagram below using a stand, ring, metal gauze, and Bunsen burner. You will use this arrangement in several experiments to heat a liquid.

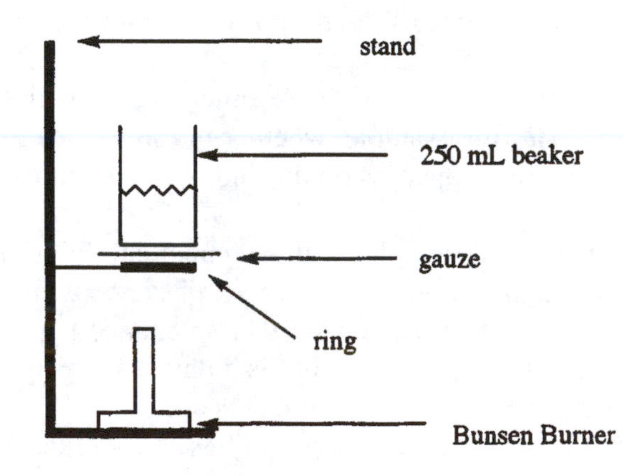

10. Half fill a 250 mL beaker with water, put a couple of boiling chips into the water and place on the gauze. This will be your "water bath."
11. Pipet about 2 mL of unknown A into a test tube, add a boiling chip, and clamp the test tube inside the beaker so that the level of A is below the level of the water. Make sure that the bottom of the test tube is not touching the bottom of the beaker.
12. Attach the tubing of the Bunsen burner to a gas nozzle.
13. Close the air holes at the base of the burner.
14. Strike a match and then open the gas tap slightly.
15. Hold the match over the top of the burner to ignite the flame.
16. Adjust the air holes until the flame is pale blue.
17. See your TA if you are at all unsure of the procedure.
18. Begin heating the water bath from below while stirring the water using your stirring rod.

CAUTION! **UNKNOWNS MAY BE FLAMMABLE. HEAT WITH CARE!**

19. Watch your liquid unknown while heating and stirring.
20. When it begins to boil, measure the temperature of the water by inserting a thermometer into the water adjacent to the test tube. Note: If the water in the water bath is close to boiling and the unknown still has not boiled, record as >90°.
21. Record this temperature (*9*).
22. Repeat once more for A and twice for B.
23. Calculate the average boiling point of each unknown (*10*).

(iv) Solubility

24. Pipet about 2 mL of A each into three test tubes.
25. To the first test tube, carefully add about 2 mL of water. Do the liquids mix or form two layers? If they mix, then A is soluble or miscible with water; otherwise it is insoluble. Record your observation (*11*).
26. To the second and third test tubes, add 2 mL of alcohol and cyclohexane, respectively. Record the solubility or not of each (*12 & 13*). Repeat steps 24–26 for unknown B.

PUT ALL WASTE CYCLOHEXANE IN ORGANIC WASTE BOTTLE!

(v) Identification

Using the data on the Table below and your observations, identify each unknown (*14*).

			Solubility		
Substance	**Density (g/mL)**	**Boiling point (°C)**	**Water**	**Alcohol**	**Cyclohexane**
water	1.00	>90	-	yes	no
ethanol	0.79	79	yes	-	yes
acetone	0.79	56	yes	yes	yes
chloroform	1.48	62	no	yes	yes
cyclohexane	0.79	81	no	yes	-
toluene	0.87	>90	no	yes	yes

B. Chemical Properties

(i) Combustion in Air

27. Pipet 1 mL of unknown A into an evaporating dish.
28. Light a splint (thin piece of wood) away from the dish.
29. Carefully (!) attempt to ignite the unknown in the evaporating dish.
30. Record your observation (*15*).
31. Repeat with unknown B.

(ii) Reaction with Sodium (to be done in the hood with TA supervision)

32. Put about one inch depth of each unknown in separate 150 mL beakers.
33. Take these to the hood and obtain a piece of sodium FROM THE TA.
34. Using the tweezers, take the piece of sodium and lay it on a paper towel.
35. Dry the sodium and quickly add it to the sample of unknown A.
36. Record your observations (*16*).
37. If your sodium did not react (it shouldn't have!), remove it, dry it and add it to the sample of B. Record your observations (*16*).
38. Return any unused sodium (there shouldn't be any!) to your TA.

SODIUM REACTS VIOLENTLY WITH WATER. DO NOT TOUCH WITH HANDS OR INGEST!

Sample Calculations

When I did the experiment, I obtained the following data (you should use more significant figures than I do)!

3. Beaker and test tube = 50.0 g
4. Volume of unknown = 1.7 mL
5. My unknown plus beaker and test tube = 51.7 g
6. Mass of unknown = (mass of beaker, test tube + unknown) – (mass of beaker + test tube)

   = 51.7 – 50.0

   = 1.7 g
7. Density = mass ÷ volume

   = 1.7 g ÷ 1.7 mL

   = 1.0 g/mL

EXPERIMENT 3

An Introduction to Chemical Reactions

INTRODUCTION

Much of chemistry is concerned with chemical reactions, in which one or more substances change chemically to one or more new substances—that is, the atoms that make up the original substance(s) become rearranged to produce the new substance(s). In this lab and the one following, you will study certain aspects of chemical reactions.

A. *Principle of Conservation of Mass*

One of the most fundamental concepts in chemistry is the Principle of Conservation of Mass. As a chemical reaction involves only the rearrangement of atoms, the total mass of the materials before the reaction (the reactants) must be the same as the total mass of the materials after the reaction (the products). In the first part of the lab, you will perform three reactions. The first will demonstrate the above principle, as both reactants and products are solids. In the second and third reactions, one of either the reactants or products is a gas. You will utilize the above principle to determine how much of the gas was used up or produced, respectively.

Reaction 1—Formation of Lead Iodide:

$$Pb(NO_3)_2(s) + 2\ KI(s) \rightarrow PbI_2(s) + 2\ KNO_3(s)$$

white white yellow white

The two white reactants (lead nitrate and potassium iodide) will react together to produce yellow (lead iodide) and white (potassium nitrate) products.

This reaction is an example of a **double displacement** or **metathesis** reaction. Each reactant consists of two parts—a **cation** (lead or potassium) and an **anion** (nitrate or iodide). During the reaction, the components "switch" so that the cation of the first becomes associated with the anion of the second, and *vice versa*.

Reaction 2—Combustion of Magnesium:

$$2\ Mg(s) + O_2(g) \rightarrow 2\ MgO(s)$$

When burned, the silvery metal, magnesium, reacts with oxygen in the air to form a grayish powder, magnesium oxide.

This is an example of a **combustion** reaction, in which a substance is burned in oxygen to form the **oxides** of the component atoms. Virtually everything in your experience that is burned is undergoing a combustion reaction. Thus, when butane (which consists of carbon and hydrogen atoms) burns, the products are "carbon di*oxide*" (CO_2) and "hydrogen *oxide*" (or water, H_2O).

Because we cannot easily weigh the amount of oxygen used up in the reaction, we have to invoke the Principle of Conservation of Mass. That is, because the combined masses of the reactants (magnesium + oxygen) has to be the same as the mass of product (magnesium oxide), if we know the masses of magnesium and magnesium oxide, we can calculate the amount of oxygen used.

Reaction 3—Dehydration of Copper Sulfate Pentahydrate:

$$CuSO_4(s) \bullet 5\ H_2O \rightarrow CuSO_4(s) + 5\ H_2O\ (l)$$

blue white

Upon heating, the blue powder ("hydrated" copper sulfate) loses water to produce the white "anhydrous" species.

This reaction is an example of a **dehydration** reaction, in which a compound loses water. The dot (•) in the formula above means that water molecules are held tightly by the copper sulfate. This reaction could also be called a **decomposition** reaction, in which one reactant "decomposes" to more than one product. As the water produced is ultimately driven off as steam, we cannot weigh how much was formed. Again, therefore, we must invoke the Principle of Conservation of Mass:

$$\text{Mass of water} = \text{Mass of blue solid} - \text{Mass of white solid}$$

B. Miscellaneous Reaction Types

In the first part of the lab, you will perform three types of reactions—metathesis, combustion, and dehydration. As a pleasurable(?) end to this week's experiment, you will perform a number of different types of reactions. Some of these types will be addressed in later experiments.

Reactions 4 and 5—Colored Complexes

$$Cu(H_2O)_4^{2+}(aq) + 4\ NH_3(aq) \rightarrow Cu(NH_3)_4^{2+}\ (aq) + 4\ H_2O(l)$$

$$Co(H_2O)_6^{2+}(aq) + 6\ NH_3(aq) \rightarrow Co(NH_3)_6^{2+}(aq) + 6\ H_2O(l)$$

These reactions, in which water molecules that surround a metal ion in solution are replaced by ammonia molecules, are called **complexation** reactions. You will encounter certain types of complexes in later labs, and many areas of everyday life rely upon such species.

Reaction 6—An Acid-Base Reaction

$$NaOH(aq) + HCl(aq) \rightarrow H_2O(l) + NaCl(aq)$$

The words **acid** and **base** should be familiar to you from various classes. The concepts of acidity and basicity, however, go far beyond pH, acid rain, *etc.* and rank among the most important in chemistry. In later experiments, you will examine acids and bases in greater depth. In this experiment, however, you will perform simple tests on an acid and a base, and then observe a special form of acid-base reaction, known as a **neutralization**.

Reactions 7 and 8—You, The Alchemist!

The alchemists of medieval times attempted to "transmute" base metals to gold. While some of the alchemists were great scientists and laid the foundations for modern experimental chemistry, others were nothing but charlatans, who conned royalty, nobles and others with fraudulent claims of success. In this part of the experiment, you will turn a copper penny into "silver" and then "gold" by chemical means! Actually, you will first coat the penny with zinc by use of a **redox** reaction, and then **amalgamate** the zinc and copper into brass by heating!

Procedure

Reaction 1—Formation of Lead Iodide

1. Weigh out (on separate pieces of weighing paper) 0.5 g of lead nitrate ($Pb(NO_3)_2$) and 0.5 g of potassium iodide (KI). Record the masses in lines (*1*) and (*2*).

LEAD SALTS ARE TOXIC—HANDLE WITH GLOVES!

2. Take a CLEAN evaporating dish and weigh it (*3*).
3. Simultaneously add the lead nitrate and potassium iodide to the evaporating dish and stir the two powders together for about five minutes.
4. Weigh the evaporating dish and products and record in line (*4*).
5. Dispose of the product in the trash can labeled "LEAD WASTE". USING GLOVES, wipe the inside of the evaporating dish thoroughly with a paper towel, and dispose of this in the "LEAD WASTE" can also.
6. Add the masses of the reactants and record the sum in line (*5*).
7. Calculate the masses of the products and record in line (*6*).
8. Calculate the mass change during the course of the reaction (*7*) and discuss your result (*8*).

Reaction 2—Combustion of Magnesium

9. Take a crucible and lid, set them on a clay triangle in hood, and heat strongly for about five minutes. Crucibles get dirty very quickly in the lab and it is essential that you burn off any impurities now rather than allow them to mess up your experiment. When done, only handle the crucible using tongs.
10. After they cool, weigh the crucible and lid together (*9*).
11. Cut about 15 cm of magnesium ribbon into small strips. This will allow the reaction to occur faster. Place the strips into the crucible, replace the lid and re-weigh (*10*).
12. Put the crucible and lid on the clay triangle, and heat gently. Lift the lid occasionally. This allows a controlled reaction with oxygen to occur. If you do not use the lid and therefore expose the magnesium completely to the air, the metal would burn with a very bright flame. This would do two things—temporarily blind anyone watching, and cause some of the metal to vaporize.
13. When it appears that all the magnesium has reacted, remove the lid and continue heating for about another minute.
14. Remove the heat, and allow the crucible to cool to room temperature. Weigh the crucible, lid and product and record in line (*11*).
15. Determine the mass of magnesium at the start (*12*).

16. Determine the mass of product magnesium oxide (*13*).
17. Calculate how much oxygen was used up (*14*) and discuss your result (*15*).

Reaction 3—Dehydration of Copper Sulfate Pentahydrate

18. Weigh a 150 mL beaker (*16*).
19. Add about 1 g of the blue copper sulfate hydrate to the beaker and record the new total mass (*17*).
20. Place the beaker on a gauze pad on a ring (as in Experiment 2), put a watch glass over the top and heat the bottom of the beaker gently.
21. You should begin to observe water droplets on the side of the beaker and the solid beginning to turn white. At this stage remove the flame, remove the watch glass and then resume heating. Continue until the solid is completely white.
22. After the beaker has cooled, weigh the beaker and white solid (*18*).
23. Determine the mass of the blue crystals at the start (*19*).
24. Determine the mass of the white crystals at the end (*20*).
25. Calculate the mass of water driven off (*21*) and discuss your result (*22*).

Reactions 4 and 5—Colored Complexes

26. Add 10 mL 0.1 M $CuCl_2$ to a 100 mL beaker.
27. Carefully add 20 mL of 0.5 M NH_3 solution in 5 mL increments.

DO NOT INHALE THE AMMONIA FUMES!

28. Record your observations (*23*).
29. Repeat steps 26 through 28 using 0.1 M $CoCl_2$ instead of $CuCl_2$ (*24*).

Reaction 6—An Acid-Base Reaction

30. Measure carefully 10 mL of 1 M HCl into a 100 mL beaker. Test with blue and red litmus paper (*25*).
31. Measure carefully 10 mL of 1 M NaOH into a 100 mL beaker. Test with blue and red litmus paper (*26*).
32. Pour the NaOH into the HCl and stir vigorously. As you stir, gently touch the side of the beaker with your finger (*27*).
33. After stirring, test with blue and red litmus (*28*).

Reactions 7 and 8—You, The Alchemist!

34. Weigh about 5 g of mossy zinc in an evaporating dish.
35. Pour enough 6 M NaOH solution into the dish to cover the zinc.
36. Place the dish on a gauze on a ring (see Experiment 2) and begin heating with a Bunsen burner.

CAUTION!

SODIUM HYDROXIDE IS CAUSTIC AND HAZARDOUS TO EYES AND SKIN. ENSURE THAT THE SOLUTION DOES NOT SPATTER.

37. Clean two pre-1980 pennies with steel wool. You should ensure that the pennies are completely shiny. (The interior of pennies made after 1980 is zinc, which will melt under the conditions used in this experiment).
38. When the solution is boiling, place the pennies on top of the zinc using tongs.
39. After the pennies are "silvered," remove the burner, and then, again using tongs, pick up the pennies and cool them in a beaker of cold water.
40. Using tongs, hold one of the pennies in the blue part of the burner flame, turning it to ensure an even heating. When the penny has turned golden, cool it in a beaker of cold water. Record your observations (*29*) and keep the pennies.

How to calculate mL from M:

0.1M → 0.1 molar in 1 L of H_2O (water)

$CuCl_2$ (molec. weight) =130

13g (0.1mol) $CuCl_2$ in 1 L water

↓ 10mL

0.13g of $CuCl_2$ in 10 mL of water

Basically, $\frac{13g\ CuCl_2}{1000\ mL\ water}$ · ____

$CoCl_2$ (Molec. weight) =139

Same procedure

Sample Calculations

When I did the experiment, I obtained the following data (you should use more significant figures than I do):

1.	Mass of $Pb(NO_3)_2$ (g)	1.1 g
2.	Mass of KI (g)	1.0 g
3.	Mass of evaporating dish (g)	63.4 g
4.	Mass of evaporating dish and product (g)	65.4 g
5.	Mass of reactants (g) (*1* + *2*)	2.1 g
6.	Mass of product (g) (*4* – *3*)	2.0 g
7.	Mass gained/lost (*6*–*5*)	-0.1 g
9.	Mass of crucible and lid (g)	6.0 g
10.	Mass of crucible, lid and magnesium (g)	6.2 g
11.	Mass of crucible, lid and magnesium oxide (g)	6.4 g
12.	Mass of magnesium at start (g) (*10* – *9*)	0.2 g
13.	Mass of magnesium oxide (product) (g) (*11* – *9*)	0.4 g
14.	Mass of oxygen used up (g) (*13* – *12*)	0.2 g
16.	Mass of beaker (g)	35.1 g
17.	Mass of beaker and blue copper sulfate (g)	36.2 g
18.	Mass of beaker and white copper sulfate (g)	35.8 g
19.	Mass of blue copper sulfate (*17* – *16*) (g)	1.1 g
20.	Mass of white copper sulfate (*18* – *16*) (g)	0.7 g
21.	Mass of water driven off (*19* – *20*) (g)	0.4 g

Copper Chemistry—A Series of Reactions

INTRODUCTION

In the last experiment, you examined the Principle of Conservation of Mass and several types of reactions. In today's experiment you are going to perform five sequential reactions in which you will transform copper metal into a variety of different copper-containing substances to eventually regenerate your copper again. In theory, you should have as much copper at the end as at the start in accordance with the Principle of Conservation of Mass. Thus, by comparing the two amounts of copper you can determine both the efficiency of the reactions as well as your lab technique.

The reactions you will carry out are (not necessarily balanced equations):

1) $Cu(s) + HNO_3(aq) \rightarrow Cu(NO_3)_2(aq) + H_2O(l) + NO_2(g)$

A **redox** reaction

2) $Cu(NO_3)_2(aq) + Na_2CO_3(s) + H_2O \rightarrow CuCO_3 \bullet Cu(OH)_2(s) + NaNO_3(aq) + CO_2(g)$

A **metathesis** reaction

In addition, any excess HNO_3 reacts with some Na_2CO_3:

2a) $HNO_3(aq) + Na_2CO_3(s) \rightarrow NaNO_3(aq) + CO_2(g) + H_2O(l)$

An **acid-base** reaction

3) $CuCO_3 \bullet Cu(OH)_2(s) + \text{heat} \rightarrow CuO(s) + H_2O(l) + CO_2(g)$

A **decomposition** reaction

4) $CuO(s) + H_2SO_4(aq) \rightarrow CuSO_4(aq) + H_2O(l)$

A **metathesis** reaction

5) $CuSO_4(aq) + Mg(s) \rightarrow MgSO_4(aq) + Cu(s)$

A **redox** or **single displacement** reaction

In addition, any excess H_2SO_4 will react with Mg:

5a) $H_2SO_4(aq) + Mg(s) \rightarrow MgSO_4(aq) + H_2(g)$

A **redox** or **single displacement** reaction

Procedure

Reaction 1

1. Weigh about 0.5 g of Cu wool and record the exact mass (*1*).
2. Place the copper in a 250 mL beaker and GO TO THE HOOD.
3. Add 3 mL of concentrated HNO_3 (nitric acid) to the copper dropwise using the pipet. REPLACE THE PIPET IN THE ACID BEAKER.

HNO_3 IS A VERY STRONG ACID SO BE VERY CAREFUL. IF YOU SPILL ANY, TELL THE TA IMMEDIATELY!

The reaction will emit brown fumes of NO_2.

NO_2 IS VERY TOXIC AND SHOULD NOT BE INHALED!

4. Carefully swirl the beaker until all of the metal has disappeared and there are no more fumes emitted.
5. Carefully add 45 mL of distilled water to the reaction mixture and then carry the beaker back to your work area on the bench. Describe the appearance of the beaker contents on the worksheet (*5*).

Reaction 2

6. Very slowly (using a spatula) add about 2 g of Na_2CO_3 to the beaker while stirring. There will be a great deal of bubbling/foaming as the CO_2 is given off. Be extremely careful not to let any of your mixture spill over the sides as you will have to start again!!
7. After all foaming has stopped and you have stirred the mixture to a homogenous consistency, test the acidity with red litmus paper. If it does not turn a purplish-blue, add some more Na_2CO_3.
8. Describe the appearance of the beaker contents on the work sheet (*6*).

Reaction 3

9. Put 30 mL of water into a 100 mL beaker, and set this aside.
10. Fold a piece of filter paper in half from top to bottom to give a half-circle, and then from side-to-side to give a pie-shaped wedge. Pull the outer-most piece open to give a cone. Place this in the filter funnel, and put this on the top of a 150 mL beaker.
11. Set up the heating arrangement from Experiment 2, using the 250 mL beaker with the blue solution (the "reaction beaker"). Heat with a LOW Bunsen flame and stir.

12. When the liquid begins to steam, add 5 mL of 3 M NaOH and continue heating and stirring until the reaction mixture is a uniform black color—this is solid copper oxide and is the desired product from the reaction.
13. Heat for a minute further.
14. Remove the "reaction beaker" (remember it's hot!) and place the 100 mL beaker from step 9 on the gauze to be heated.
15. Carefully pour a small amount of the contents of the 250 mL reaction beaker onto the filter paper (which is in the filter funnel on top of the 150 mL beaker). When most of the liquid has run through, carefully add the rest of the reaction mixture, bit by bit. BE SURE THAT YOU DO NOT OVERFLOW THE FILTER PAPER.
16. When you have filtered all of the mixture from the 250 mL beaker, check the filtrate (*i. e.*, the post filtration liquid) in the receiving beaker. If this is cloudy, carefully remove the filter funnel, pour the filtrate into the 250 mL beaker and re-filter.
17. Once your filtrate is clear, turn off the Bunsen and pour about half the hot water in the 100 mL beaker into the 250 mL beaker.
18. Swirl the water around, making sure that any residual solid in the beaker is "caught" by the water, and carefully pour the water into the filter funnel (again, taking care not to overflow). This process is called "washing".
19. Once most of the water has run through the filter paper, repeat the washing process with the rest of the hot water.
20. Keep the solid in the filter paper and funnel for the next part and discard the filtrate down the sink drain with plenty of water.

Reaction 4

21. Place the filter funnel on top of the 250 mL beaker. You are about to "do a reaction" in the filter funnel.
22. Add 20 mL of 3 M H_2SO_4 carefully in portions to the solid. The mixture should bubble and possibly emit some fumes (do not inhale!!), and a translucent, colored liquid should drip into the beaker. Describe the appearance of this liquid on the Data Sheet (7).
23. When all of the liquid has dripped through, your solid should have disappeared. If you have any solid left, move the filter funnel to on top of a CLEAN 100 mL beaker, and run the colored filtrate from the 250 mL beaker back through the filter. Repeat this process until no solid remains on the filter paper.
24. Save the colored filtrate for the next part.

Reaction 5

ENSURE THAT NEITHER YOU NOR YOUR NEIGHBORS HAVE A BUNSEN BURNER GOING WHILE YOU ARE DOING THIS.

25. Take 1.5 g of Mg pieces, and, piece by piece, drop this into the beaker, stirring as you do it. A solid should start to appear. Describe the appearance of the solid on the data sheet (*8*).
26. Add Mg until the last piece does not react and then add about 5 mL of 3 M H_2SO_4 to dissolve the excess magnesium.
27. Write your name on and then weigh a new piece of filter paper and record this mass (*2*).
28. Filter the mixture, and wash twice with 10 mL each of water.
29. Very carefully, remove the filter paper and place it in the oven.
30. Your TA will weigh the dried paper and product and post the results outside your lab room. Be sure to come up and record your result within a couple of days (*3*).
31. Calculate the mass of your product (*4*) and compare it to the amount with which you began.

Preparation and Properties of Gases

Introduction

In today's experiment, you will prepare a number of simple gases and examine some of their properties. The objectives of this experiment are threefold:

(a) to familiarize you with gas production and handling techniques,

(b) to convince you that gases really exist,

(c) to learn the common tests for these gases.

Procedure

You will use two different methods of producing gases. In both cases, a gas forming reaction takes place in a vessel that is sealed except for a hole in the top. The gas produced escapes through the hole, along a piece of rubber tubing, and into what is usually an inverted, water or air-filled container. The gas displaces the water or air in the container.

The difference between the two gas production methods arises from the difference in reactivity. Three of the reactions you will do are spontaneous at room temperature, while two need to be heated.

The general procedures are listed below. For each of the following gas productions, only the indicated steps of the particular method are to be changed.

METHOD I—Spontaneous reactions (DONE ON BENCHTOP), gas collected OVER WATER

1. Put one reactant into a 250 mL Erlenmeyer flask.
2. If not already done, take the stopper which has the right-angle glass tubing in one hole and attach a length of rubber tubing to the end of the glass tubing.
3. Fill a water trough half full of water.
4. Fill the appropriate number of 250 mL Erlenmeyer flasks full of water.
5. Take a flask, cover its mouth with your hand, and invert it into the trough. If necessary, do the same thing with any remaining flasks.
6. Place the free end of the rubber tubing under the water in the water trough.
7. Pour the second reactant into the flask and put the stopper in the flask.

8. Bubbles should start coming out of the end of the tubing in the water trough.
9. Add the second reactant as needed to keep the bubbles forming at a moderate rate.
10. After the bubbles have been going for 30 seconds, place the end of the tubing under the mouth of one of the flasks.
11. The gas will displace the water and fill the flask. Once the flask is full of gas, seal it with a stopper and remove it from the trough.
12. Repeat this process to fill any remaining flasks.

METHOD II—Heated Reactions (DONE IN HOOD), gas collected by displacement of AIR

1. Place reactants in a 250 mL Erlenmeyer flask.
2. Place a one-hole stopper with right-angled tubing in the flask.
3. Set up a stand with a ring, wire gauze and a clamp to hold the 250 mL Erlenmeyer flask.
4. Place flask on ring stand and clamp it so that it will be stable.
5. Clamp two Erlenmeyer flasks vertically.
6. Fill a 1 liter beaker with 800 mL of cold tap water.
7. Gently heat the contents of the reaction flask. The solution should begin to bubble. If the bubbling becomes too vigorous, the reaction may be slowed by placing the flask in the liter beaker of water.
8. Gas should be coming out of the end of the tubing. Hold a piece of damp litmus paper over the end of the tubing. When this turns color, insert the tubing all the way to the bottom of one of the collection flasks.
9. Hold a piece of damp litmus paper at the mouth of the collection flask. When this turns color, remove tubing and place in next flask. Stopper the original flask.
10. When both flasks are filled turn off the heat. After the reaction flask has cooled, fill with water and pour down the sink drain.

You might find it easier to tear this page out, and use it for reference rather than having to keep turning back all the time. Record any observations from each experiment on the data sheet.

Gas 1 Carbon Dioxide

(i) Preparation: This procedure uses METHOD I

<u>Reaction:</u> $2H^{+}(aq) + CO_3^{2-}(aq) \rightarrow CO_2(g) + H_2O(l)$

New Step 1: Place 20 g of $CaCO_3$ chips into one flask and add 50 mL of water.

New Step 4: Fill three 250 mL Erlenmeyer flasks with water.

New Step 7: Pour 2 mL of 6 M HCl into the reaction flask and quickly stopper.

HCl CAUSES SEVERE BURNS

New Step 9: Add HCl as needed.

(ii) Properties

1. Light a wooden splint and insert it into one of the flasks of gas. Record observations (*2*).
2. Place about 2 mL of acetone in an evaporating dish. Very carefully, ignite with a match. Hold a second Erlenmeyer flask about 4 inches over the flame and "pour" the gas over it. Record your observations (*3*).
3. Pour 25 mL of saturated $Ca(OH)_2$ solution (limewater) into the third flask and swirl. Record your observation (*4*).

Gas 2 Hydrogen

(i) Preparation: This procedure uses METHOD I

Reaction: $2\ HCl(aq) + Zn(s) \rightarrow H_2(g) + ZnCl_2(aq)$

New Step 1: Place 10 g of mossy zinc into the flask and add 37.5 mL of water.

New Step 4: Fill one 250 mL Erlenmeyer flask with water.

New Step 7: Pour 2 mL of 6 M HCl into the reaction flask and quickly stopper.

New Step 9: Add HCl as necessary to keep the bubbles forming.

(ii) Properties

1. Light a splint.
2. Unstopper the flask and hold splint to its mouth. Record observation (*6*).

Gas 3 Oxygen

(i) Preparation: This procedure uses METHOD I

Reaction: $5\ H_2O_2(aq) + 2\ MnO_4^-(aq) + 6\ H^+(aq) \rightarrow 5\ O_2(g) + 2\ Mn^{2+}(aq) + 8\ H_2O(l)$

New Step 1: Pour 100 mL of 3% H_2O_2 and 60 mL of 3 M H_2SO_4 into the flask.

New Step 4: Fill two 250 mL Erlenmeyer flasks with water.

New Step 7: Add 1.5 g of $KMnO_4$ into the reaction flask and quickly stopper.

$KMnO_4$ IS EXPLOSIVE IN CONTACT WITH ORGANICS H_2SO_4 IS A VERY CORROSIVE ACID AND SHOULD BE HANDLED WITH CARE. REPORT ANY SPILL TO THE TA IMMEDIATELY.

New Step 9: Add $KMnO_4$ as necessary.

(ii) Properties

1. Light a splint, then extinguish the flame until only glowing embers are left. Insert into one flask and record your observations (*8*).
2. Light a Bunsen, and, using tongs, hold a small wad of steel wool in the flame until it glows. Insert it into the second flask, observe and record (*9*).

CUT STEEL WOOL WITH SCISSORS NOT HANDS. HOLD TONGS OR SPLINT WELL AWAY FROM GLOWING END AS COMBUSTION MAY BE VIGOROUS.

DO THE REMAINING EXPERIMENTS IN THE HOOD.

Gas 4 Hydrogen Chloride

(i) Preparation This procedure uses METHOD II.

<u>Reaction:</u> $H_2SO_4(aq) + NaCl(aq) \rightarrow HCl(g) + Na^+(aq) + HSO_4^-(aq)$

New Step 1: Place 2.5 g of NaCl and 10 mL of conc. H_2SO_4 into the flask.

H_2SO_4 IS A VERY STRONG ACID AND DEHYDRATING AGENT. REPORT ANY SPILLS TO TA IMMEDIATELY!

New Step 5: Clamp two Erlenmeyer flasks upright.

Steps 8, 9: Use BLUE LITMUS, WHICH WILL CHANGE COLOR TO RED.

HCl GAS IS EXTREMELY IRRITATING TO EYES AND LUNGS. DO NOT INHALE OR TOUCH.

(ii) Properties

1. Unstopper a flask and quickly add 10 mL of water. Restopper and shake vigorously. Test the water with blue litmus paper. Record and explain your observation (*11*).
2. Save the second flask, because it will be used in the next part.

Gas 5 Ammonia Use Z-shape(?)

(i) Preparation: This procedure uses METHOD II.

Reaction: 2 $NH_4Cl(aq)$ + $Ca(OH)_2(aq) \rightarrow$ 2 NH_3 (g) + $CaCl_2(g)$ + 2 $H_2O(l)$

New Step 1: Place 5 g of NH_4Cl, 5 g of $Ca(OH)_2$ and 5 mL of water into the flask.

New Step 5: Clamp two Erlenmeyer flasks UPSIDE DOWN.

Steps 8, 9: Use moist red litmus (which will turn blue at the appropriate time).

NH_3 IS AN IRRITANT GAS. DO NOT INHALE!

(ii) Properties

1. "React" the contents of one flask with water as for HCl using red litmus. Record and explain (*13*).
2. Take the remaining flask of HCl and the second of NH_3. Wearing gloves, open the stoppers simultaneously and hold the mouths close together. Record your observation (*14*).

Gas Laws

Introduction

The properties of gases have long fascinated chemists, and indeed, were investigated extensively to derive many of the fundamental chemical theories. In today's experiment, you will investigate three of the so-called "gas laws"—Boyle's, Charles' and Avogadro's law. In the post-lab questions, you will then combine these into the "ideal gas law." In addition, you will examine diffusion of gases through air; thereby deriving Graham's law.

The first two laws concern relationships between the three basic physical parameters of gases: volume, pressure, and temperature.

Volume of a gas is simply the volume of the vessel containing the gas, as any gas will "fill" its entire container. We use liters as the unit.

Pressure is defined as the force per area caused by the gas on the walls of its container.

Temperature is obvious. When dealing with gases, we use the unit **Kelvin** where the temperature in Kelvin = (the temperature in °C + 273).

Boyles' Law concerns the relationship between volume and pressure, and Charles' Law concerns that between volume and temperature.

Avogadro's law is more of a chemistry thing, relating amount of a gas (in moles) to its volume.

The final law, that of Graham, concerns the rate of diffusion of a gas. Diffusion is the process in which a gas passes through another under constant pressure. The "rate of diffusion", therefore, is related to the velocity of gas molecules.

All four laws should be familiar to you from class. In this lab, you should become convinced that class is correct!

Procedure

A. Boyle's Law

Ordinarily, one measures pressure using a manometer and units of mm of mercury (mm Hg). Unfortunately, the thought of so many liters of mercury in a general chemistry lab filled the safety office with fear. Accordingly, we will mimic pressure using textbooks.

1. If it has not already been done for you, draw 30 mL of air into a 50 mL plastic syringe, then seal the end with a syringe tip. Record the volume (*1*).

2. Clamp the syringe vertically with the tip pointing downward.
3. Balance a textbook on the plunger, gently press down and then allow the plunger to spring up. Record the volume of air (*2*).
4. Balance a second textbook on the plunger and repeat step 3. Continue with 3 and 4 textbooks, recording the volume of air (*3 – 5*).
5. Plot a graph of number of books (on the horizontal axis) vs. 1/volume (on vertical axis) and determine the relationship between number of books and volume (*6*).

B. Charles' Law

6. If it has not already been done for you, draw 6 mL of air into a 10 mL plastic syringe, then seal the end with a syringe tip.
7. You will use the following three baths that will have been set up for you:
   a. A hot bath. Use either a water bath in the lab or set up a beaker of hot water.
   b. An ice bath. Fill a second 250 mL beaker with ice and water.
   c. An alcohol bath. Use an isopropyl alcohol/dry ice bath.
8. Measure the temperature of the air by placing the thermometer next to the syringe, and record it and the volume of air in the syringe (*7*).
9. Place the syringe in the hot bath and record the temperature and volume after about 2 minutes (*8*).
10. Put the syringe in the ice bath and record the temperature and volume after about 2 minutes (9).
11. Place the syringe in the alcohol bath and record the temperature (using the alcohol thermometer provided) and volume after about 2 minutes (*10*).
12. Plot a graph of volume (on y axis) versus temperature (on x axis) and determine the relationship between temperature and volume (*11*).

C. Avogadro's Law

In a previous lab, you carried out the reaction of zinc with hydrochloric acid, making hydrogen. In this section of today's experiment, you will repeat this reaction, measuring the volume of hydrogen produced by a known amount of zinc.

13. Weigh a large test tube (*12*).
14. Add approximately 0.05 g of zinc powder and record the mass (*13*).
15. Clamp the tube vertically, insert a one-hole stopper that has a right angled piece of glass tubing in the hole. Attach a piece of rubber tubing to the end of the glass tubing if not already done.
16. Clamp a 50 mL glass syringe horizontally and attach the nozzle to the end of the piece of rubber tubing in step 15. A diagram of the final apparatus is shown over the page:

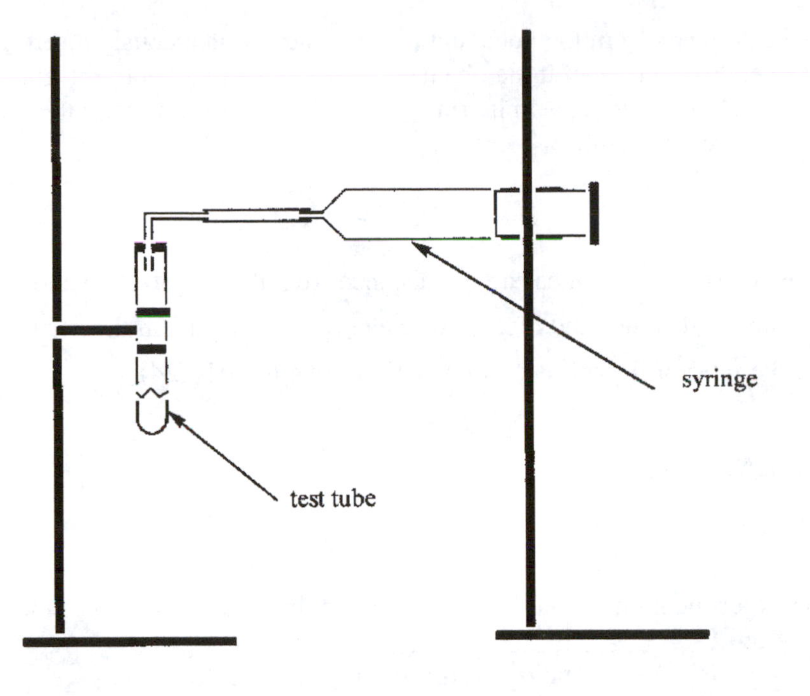

17. Push the plunger of the syringe all the way in, and record the initial volume reading (*16*).
18. Open the stopper and add approximately 5 mL of 3 M HCl (a vast excess). Immediately replace the stopper.
19. Allow the reaction to proceed to completion, rotating (without pulling) the plunger to prevent it from sticking. When the plunger has not moved for a minute, record the volume (*17*).
20. Empty the apparatus and repeat steps 13 through 19 twice more. For the second attempt, use approximately 0.1 g of zinc and for the third, approximately 0.15 g.
21. Plot a graph of volume (on y) against number of moles of zinc (on x) and determine the relationship between number of moles and volume (*19*).
22. Measure the temperature (*20*) and pressure (*21*) of the surroundings.

D. *Graham's Law of Diffusion*

23. Place a ball of cotton wool in the hole of each of two partially holed stoppers.
24. Clamp a 1 meter piece of glass tubing horizontally and label one end HCl, the other NH_3.

IN THE HOOD:

25. Wearing gloves, pipet several drops of concentrated HCl onto the cotton in one stopper then pipet several drops of concentrated NH_4OH onto the cotton in the other stopper.

HCl AND NH_3 CAN CAUSE BURNS AND IRRITATE EYES AND LUNGS. USE EXTREME CAUTION!

26. Remove the stoppers from the hood and quickly and simultaneously insert the stoppers into the appropriate ends of the tube. It may be necessary to tape them onto the glass tube so that they do not come out. After a while, a white ring with NH_4Cl will form along the tube, according to the equation (see previous lab):

$$NH_{3(g)} + HCl_{(g)} \rightarrow NH_4Cl_{(s)}$$

27. As soon as the ring forms, measure the distance from the ring to each end of the tube (*22, 23*).
28. Rinse the tube with water and dispose of the cotton wool balls in the basin in the hood.
29. Calculate the ratio of the diffusion rate of HCl : rate of NH_3 (*24*).

Sample Calculations

A. *Boyle's Law*

When I did the experiment on a small scale, I obtained the data in the Table below:

# Books	Volume (mL)	1/V (1/mL)
0	6.0	0.1666
1	4.0	0.250
2	3.0	0.333
3	2.4	0.417
4	2.0	0.5000

A plot of 1/V vs. # books looks like:

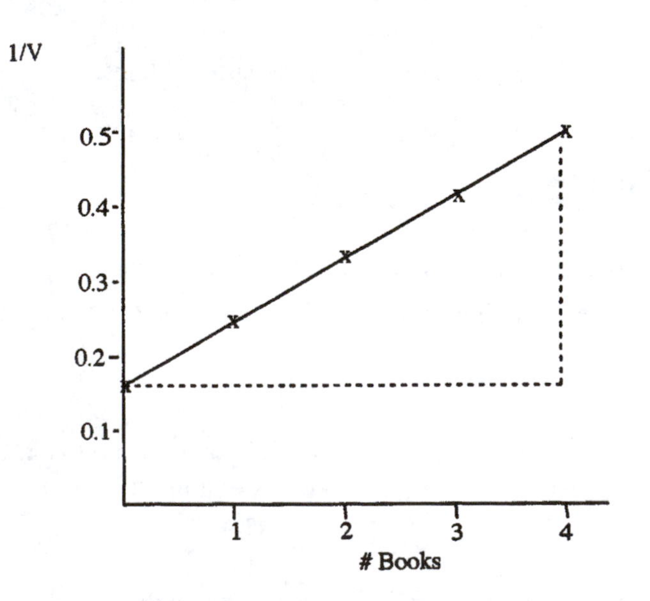

As this is a straight line, we can put it in the form y = mx + c,

or, 1/V = (m x # books) + c

Take the slope as shown, m = change in 1/V ÷ (change in # books) = 0.33 ÷ 4 = 0.08

Substitution into the equation 1/V = (0.08 x # books) + c, using the data for 0 books gives c = 0.17

Thus, the relationship between V and "P" is:

$$1/V = 0.08\ P + 0.17$$

B. Charles' Law

When I did the experiment, I obtained:

	Temp (˚C)	Vol (mL)	T (K)
Room temp	22	6.0	295
Hot bath	62	6.8	335
Ice bath	2	5.6	275
Alcohol bath	-22	5.3	251

A plot of V vs T looks like:

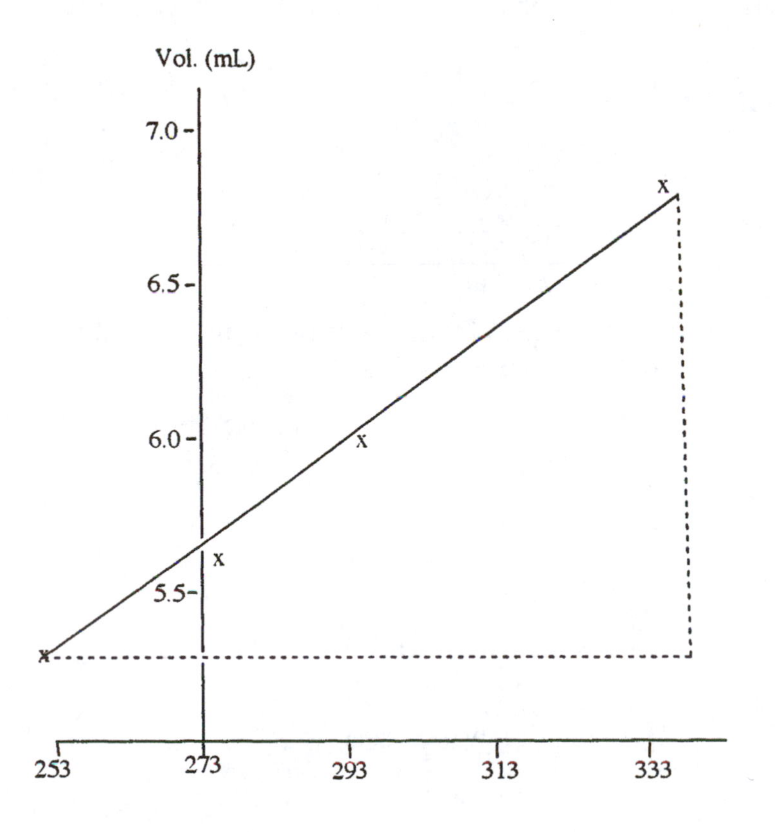

Again, as this is a straight line, its equation can be expressed in the form:

$$V = mT + c$$

Taking the slope as shown:

$$m = (\text{change in V}) \div (\text{change in T}) = 1.5 \div 84 = 0.018$$

$$\text{so: } V = 0.018\ T + c$$

Substituting in the room temperature data gives:

$$6 = (0.018 \times 295) + c$$

$$6 = 5.31 + c$$

$$c = 0.69$$

Thus, $V = 0.018\ T + 0.69$

C. Avogadro's Law

When I did the experiment, I obtained:

14.	Mass of zinc (*13 – 12*) (g)	0.051	0.110	0.156
15.	Moles of zinc (*14*/65.4)	7.80×10^{-4}	1.68×10^{-3}	2.39×10^{-3}
18.	Volume of hydrogen (*17 –16*) (L)	9.4 mL	20.0 mL	28.8 mL

When plotted out as a straight line, we get the relationship, V (L) = number of moles × 12.0

D. Graham's Law

When I did this experiment:

22. Distance of HCl from origin = 0.42 m
23. Distance of NH_3 from origin = 0.58 m

Now rate of diffusion is essentially the same as velocity, and, rate = distance ÷ time. In the same amount of time, the HCl traveled 0.42 m and the NH_3 traveled 0.58 m.

Thus, for HCl:

$$\text{rate (HCl)} = 0.42\ \text{m} \div \text{time}$$

or

$$\text{time} = 0.42 \div \text{rate (HCl)}$$

For NH_3:

$$\text{rate } (NH_3) = 0.58\ \text{m} \div \text{time},$$

or

$$\text{time} = 0.58\ \text{m} \div \text{rate } (NH_3)$$

Since travel times are equal, we can set these equations equal to each other:

$$0.42 \div \text{rate (HCl)} = 0.58 \div \text{rate } (NH_3)$$

or

$$\text{rate (HCl)} \div \text{rate } (NH_3) = 0.42 \div 0.58 = 0.72$$

According to Graham:

$$\text{rate (HCl)} \div \text{rate } (NH_3) = \sqrt{\,[\text{molecular mass } (NH_3) \div \text{molecular mass (HCl)}]}$$

or

$$(0.72)^2 = \text{molecular mass } (NH_3) \div \text{molecular mass (HCl)}$$

Use this expression (with your data instead of mine!) for the post lab.

Name Shahanna Wade TA Yen Young

DATA SHEET—EXPERIMENT 6

Gas Laws

A. Boyle's Law

	# books	volume of air	1/V
1.	0	______	______
2.	1	______	______
3.	2	______	______
4.	3	______	______
5.	4	______	______

6. What is the relationship between # books and volume of air? (Write the equation)

B. Charles' Law

		Temp.	Vol.
7.	Room temperature	______	______
8.	Hot bath	______	______
9.	Ice bath	______	______
10.	Alcohol bath	______	______

11. What is the relationship between volume and temperature? (Write the equation)

C. Avogadro's Law

12. Mass of flask (g) ______ ______ ______
13. Mass of flask + zinc (g) ______ ______ ______
14. Mass of zinc (*13 –12*) (g) ______ ______ ______
15. Moles of zinc (*14*/65.4) ______ ______ ______
16. Initial volume (L) ______ ______ ______
17. Final volume (L) ______ ______ ______
18. Volume of hydrogen (*17–16*) (L) ______ ______ ______
19. Relationship between volume and number of moles (Write the equation)

20. Temperature of surroundings (K) ______
21. Atmospheric pressure (atm) ______

D. Graham's Law of Diffusion

22. Distance of HCl from ring ______

23. Distance of NH_3 from ring ______

24. Ratio of rates HCl ÷ NH_3 ______

POSTLAB

1. According to the gas law, PV = nRT, there is no " + c" in Boyles' Law. To what is this extra constant due?

2. Using Charles' Law data, calculate your value of absolute zero, in K, given that the volume at this temperature = zero.

3. Using your Avogadro's Law data, calculate the value of R.

4. What should your ratio of rates have been for the Graham's Law experiment? Account for the difference.

Determination of Molar Mass by the Dumas Method

INTRODUCTION

One of the most critical concepts in chemistry is that of molar mass (*i e*., the mass of one mole of a substance). Without a full understanding of this idea, control of stoichiometry and therefore of reactions in general is impossible.

These days, the determination of the molar mass of a compound is usually accomplished by some form of electronic instrumentation—either directly (using mass spectrometry) or indirectly (using, for example, the molecular formula obtained by X-ray crystallography). In the early days of chemistry, however, other methods had to be used. In this experiment, you will utilize one of the original methods to determine molar mass.

The method used in today's experiment is based on the gas laws. If one examines the Ideal Gas Law (equation 1), it is readily seen that it contains one constant (R), three experimentally measurable parameters (V, P and T) and one other parameter:

$$PV = nRT \tag{1}$$

where,

P = the pressure (in atmospheres), V = the volume of the sample (in liters), n = the number of moles of the sample, R = 0.082 atm L/mole K, T = the temperature (in Kelvin)

Equation 1 can be rearranged to isolate the unmeasurable parameter (n);

$$n = PV \div RT \tag{2}$$

Thus, it is possible to determine the number of moles in a gas sample by simply measuring the volume, pressure and temperature of the sample.

So what? Well, the number of moles in any sample of any pure substance is related to the mass of the sample in grams (g) and the molar mass (MM):

$$n = g \div MM \tag{3}$$

Substitution of equation 3 into equation 2 yields,

$$g \div MM = PV \div RT \qquad (4)$$

which rearranges to,

$$MM = gRT \div PV \qquad (5)$$

Every variable on the right hand side of equation 5 may be measured in a laboratory, thus allowing the determination of the molar mass.

In this experiment, you will boil an unknown liquid and, by determining the volume, pressure, temperature and mass of the resultant vapor, calculate the molar mass of the liquid.

PROCEDURE

1. Obtain a clean, dry 125 mL Erlenmeyer flask and weigh it (*1*).
2. Add about 3 mL of your unknown liquid to the flask, cover the mouth of the flask with a piece of aluminum foil and fasten the foil in place with an elastic band. Pierce a few holes in the foil using a pin.
3. Clamp the Erlenmeyer flask in a water bath so that the water level in the bath is well above the liquid in your flask.
4. You should begin to see your unknown liquid begin to vaporize, and some of the vapor will escape through the holes in the top (this can be more easily seen if you hold a piece of white paper behind the vapor). Continue heating until there is no more liquid left.
5. Measure the temperature of the water bath by putting the thermometer in the water bath as close as possible to the flask without touching it (*3*). Remove the flask and allow it to cool to room temperature. At this point, you could begin the second trial (step 7).
6. Dry the outside of the flask, go to the balance, **remove the foil and band,** and weigh (*2*).
7. Repeat the entire procedure.
8. Two final parameters are needed—the volume of the gas and the pressure. The pressure of the vapor is the same as the atmospheric pressure, as you allowed the vapor to escape until the pressure inside the flask was the same as that outside. Ask your TA for the current atmospheric pressure (*4*).
9. The volume of the vapor is the same as that of the flask. To determine this, fill the flask to the top with water. Measure the volume of the water using a 50 mL cylinder (you will have to fill the cylinder from the flask about three times). Record in line (*5*).
10. Perform your calculations and give your data to the TA.
11. At the end of lab, the TA will post all the class data outside the lab room. At some time during the next week, copy this data (*11*) and calculate the class mean (*12*) and standard deviation (*13*).

Sample Calculations

When I did the experiment using acetone as my unknown (C_2H_6O), I obtained the following results:

1. Mass of flask = 74.99 g
2. Mass of flask and vapor = 75.22 g
3. Temperature = 97 °C
4. Pressure = 0.98 atm
5. Volume = 145 mL
6. Mass of vapor is simply = (mass of flask and vapor) - (mass of flask)
   = 75.22 - 74.99 g
   = 0.23 g
7. Volume of flask (L) = volume (mL) ÷ 1000
   = 0.145 L
8. Temperature in K = temp in °C + 273
   = 97 + 273
   = 370 K
9. Molar mass in g/mole = gRT ÷ PV
   = (0.23 g x 0.082 x 370 K) ÷ (0.98 atm x 0.145 L)
   = 49.1 g/mole

Mean and Standard Deviation

Throughout the rest of the two semester sequence, you will use proper methods for consideration of your data. To do this, we need to think about some statistical concepts.

Most importantly—a real scientist NEVER bases anything on one result. There are many reasons for this—a bad experiment, a freak occurence, etc. Therefore, to be sure of a result, a scientist should do an experiment many times. Obviously, this would get boring and be time-consuming for a laboratory course, so you will do any experiment a maximum of twice and then combine your data with that of the rest of the class.

When you do consider a lot of data for one experiment, if you plot a graph of how many times a particular data point occurs, you should see a bell curve or Gaussian distribution. For example:

In investigating a similar experiment, a class obtained the following values for their molar mass:

42.1	44.6	43.9	44.2	41.0	47.2	88.9	40.2	44.0	43.2	48.7	39.3
42.2	43.4	44.7	44.5	45.7	46.2	45.2	44.9	44.1	45.8	44.2	45.7

If we plot this data as explained above:

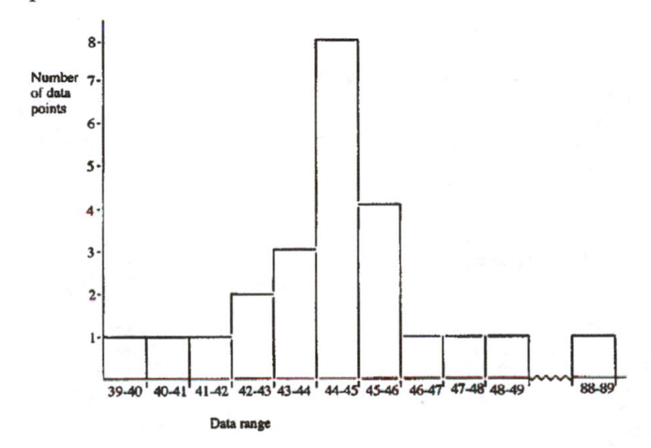

The first thing to note is that one data point is way off. As it is so bad in comparison to the close similarity of the other twenty three, it is a reasonable assumption that this particular trial was bad, and the point can be ignored. You should note that, had we done only one or two trials, we could not do this and it would adversely affect our conclusions.

Ignoring that one point, the rest of the results form a nice bell curve. We can analyze this to give two numbers—the mean (or average) and the standard deviation (or average imprecision).

i) The Mean

To get the mean: [mean = $\Sigma(x) \div n$]

a) Add up all the data points to be considered.

b) Divide by the total number of data points used.

Thus, in our example:

The sum of all the data = 1015

We used 23 data points (ignoring the 88.9)

Thus the mean value is 1015 ÷ 23 = 44.1

ii) The Standard Deviation (Standard deviation = $\sqrt{[\Sigma(x-\bar{x})^2 \div (n-1)]}$

a) Subtract the mean from each data point.

b) Square each of the numbers obtained in (a).

c) Add up all the numbers obtained in (b).

d) Divide the result of (c) by (the number of data – 1).

e) Take the square root to get the standard deviation.

It is often easier to do this in tabular form:

Data Point	(a) = data point – 44.1	(b) = (a)^2
42.1	-2.0	4.0
44.6	0.5	0.25
43.9	-0.2	0.04
44.2	0.1	0.01
41.0	-3.1	9.61
47.2	3.1	9.61
40.2	-3.9	15.21
44.0	-0.1	0.01
43.2	-0.9	0.81
48.7	4.6	21.16
39.3	-4.8	23.04
42.2	-1.9	3.61
43.4	-0.7	0.49
44.7	0.6	0.36
44.5	0.4	0.16
45.7	1.6	2.56
46.2	2.1	4.41
45.2	1.1	1.21
44.9	0.8	0.64
44.1	0	0
45.8	1.7	2.89
44.2	0.1	0.01
45.7	1.6	2.56

(c) = Sum of all of parts (b) = 102.65

(d) = answer in (c) divided by (23 – 1) = 4.67

(e) = $\sqrt{}$ (d) = 2.16

Thus, our final data based on the class result would be that the molar mass is:

$$44.1 \pm 2.2$$

At this point, you can compare your results to the overall, and decide what might have caused any discrepancy (*e. g.*, lost product, bad technique, *etc*).

Thermochemistry

INTRODUCTION

The understanding of chemical and physical changes may be greatly enhanced by an understanding of the energy changes involved. In order to measure these energy changes, we make use of the First Law of Thermodynamics, which states that energy may be transferred but never created or destroyed (equation 1). Thus, the energy gained by a process under study must be lost by the surroundings, and *vice-versa*. By measuring the energy changes of the "surroundings", we may derive the energy changes of the process.

$$\text{heat change of the reaction} = \text{-heat change of the surroundings} \qquad (1)$$

Such energy changes may take many forms, of which heat, sound, light and mechanical are the most common. In today's lab, we will measure the change in heat for a number of processes.

The heat change (Q calories) of an object is defined using three parameters: the mass (m grams) of the object, the temperature change (ΔT °C), and the specific heat, c, in calories per gram per °C (which is defined as the amount of heat needed to raise the temperature of 1 gram of the object by 1 °C):

$$Q = m \times c \times \Delta T \qquad (2)$$

This equation on its own is of little use. However, when combined with the First Law of Thermodynamics, we obtain the valuable relationship (3):

$$Q_{reaction} = -Q_{surroundings} \qquad (la)$$

$$m_{reaction} \times c_{reaction} \times \Delta T_{reaction} = -m_{surroundings} \times c_{surroundings} \times \Delta T_{surroundings} \qquad (3)$$

In all of our studies today, the "surroundings" will be WATER. This choice is made for two reasons:

(1) water is easily obtainable and inexpensive, and

(2) the specific heat of water is 1 calorie per gram per degree centigrade, which makes calculations much easier (those of you used to using joules should appreciate this!)

In each investigation, we will measure the heat change of an amount of water contained in an insulated "calorimeter" (actually, it's a coffee cup) and relate that heat change to the heat change of interest. Obviously, the calorimeter itself is part of the surroundings, so this must be incorporated into the calculation. Thus,

$$Q_{process} = -(Q_{water} + \text{heat absorbed by calorimeter}) \qquad (4)$$

Before beginning our experiments, therefore, we must measure the mass and specific heat of the calorimeter.

Having done this, the three investigations we will perform are as follows:

a) measure the specific heat of a metal,

b) measure the heat of solution of three salts, and

c) measure the energy change of a reaction.

PROCEDURE

A. Calorimeter Measurement

Obtain two coffee cups (one inside the other), a stirring rod, thermometer, stand and clamp. These constitute your "calorimeter", which should be set up as shown in the diagram. The object of the first part of the experiment is to measure how much heat this apparatus absorbs.

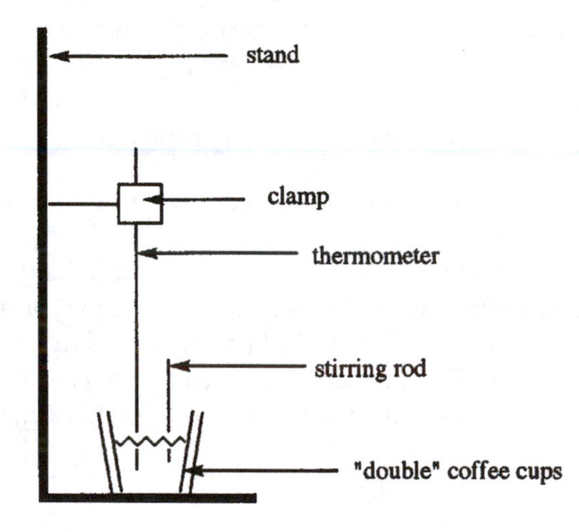

1. Add 50.0 mL of water to your coffee cup.
2. Take 50.0 mL of water in a separate beaker and heat to about 60 °C. When the "hot water" is ready, measure the temperature of the water and record (*1*).
3. Now measure the temperature of the water in the calorimeter, and record (*2*).
4. Add the hot water to the calorimeter, and mix thoroughly, while watching the thermometer. Record the maximum temperature of the mixture (*3*).
5. Calculate the (mass $\times$ c) or **heat capacity** of the calorimeter (*4*).

B. Specific Heat of a Metal

Now that our apparatus is calibrated, we can start to do some real experiments. The first experiment is to measure the specific heat of an unknown metal.

6. Obtain 30 g of the unknown metal, weigh it and record the mass in line (*5*).
7. Set up a water bath as in experiment 2, heat the water until boiling, and then carefully put your metal into it. Keep boiling for another 8–10 minutes, to allow the metal to reach the same temperature as the water. Try to avoid heating the metal directly!

8. While this is being done, measure 30.0 mL of water and add to the calorimeter. Now set up the calorimeter as before. Measure the temperature of the water in the calorimeter (*6*).
9. Once the metal has reached "thermal equilibrium" with the hot water, measure the temperature of the hot water (*7*), and quickly add the metal to the calorimeter.
10. Stir the water in the calorimeter, and record the temperature every 10 seconds (filling in the chart (*8*) on the data sheet). Obviously, the temperature will increase at the start. Continue until the temperature of the water has reached a maximum and then cooled down by about 5 degrees.
11. Plot the data from lines (*6*) and (*8*) with time on the x axis and temperature on the y axis.
12. Draw 2 sloped lines through the points where temperature is rising, and then falling down. The intersect of these lines is the maximum temperature (*9*).
13. Calculate the specific heat of the metal (*10*).

C. Molar Heat of Solution

The molar heat of solution of a substance is the amount of energy involved in dissolving one mole of the substance in a solvent (usually water). In this part of the experiment, we will measure the heat of solution of three ionic compounds: sodium chloride (NaCl), ammonium chloride (NH_4Cl), and sodium hydroxide (NaOH).

You should note that there is no such thing as the mass or specific heat of a reaction, which simplifies the calculations.

The procedure is the same in each case, so it will be detailed for only one of the three.

14. Set up the calorimeter with 25.0 mL of water. Measure the temperature (*11*).
15. Weigh about 1.5 g of your solid and record the mass (*12*).
16. Quickly add the solid into the calorimeter, stir and watch the thermometer. Record either the maximum (if the temperature goes up) or the minimum value (if the temperature goes down) on line *13*.
17. Calculate the heat of solution (*17*).

NaOH IS HIGHLY CAUSTIC. DO NOT TOUCH—IF YOU DO, WASH OFF IMMEDIATELY.

CAUTION!

NH_4Cl RELEASES FUMES OF AMMONIA. DO NOT INHALE!

D. Energy of Reaction

In this part of the lab, we are going to measure the energy change for the following reaction:

$$HCl + NaOH \rightarrow NaCl + H_2O$$

The energy change for a reaction is usually expressed PER MOLE, so this will also have to be incorporated into our calculation. However, as in Part C, one never talks about the mass or specific heat of a reaction, so this again simplifies the calculations considerably.

18. Place 52.0 mL of 1.0 M NaOH into the calorimeter and pour 50.0 mL of 1.0 M HCl into a beaker.
19. Place the two side-by-side until they attain the same temperature. Record this value (*18*).
20. Add the HCl quickly to the calorimeter and record the temperature when it reaches a maximum (*19*).
21. Calculate the molar heat of reaction (*21*).

Sample Calculations

A. Calorimeter Measurement

In this case, the hot water is our "reaction", and the cold water and calorimeter constitute our "surroundings". As the "reaction" cooled down, it lost energy, which was transferred to the surroundings, which therefore gained energy:

$$\text{heat change of hot water} = -\text{heat change of surroundings} \qquad (5)$$

When I did this experiment, my data were:

1. Temperature of hot water = 62 °C
2. Temperature of cold water = 22 °C
3. Temperature of mixture = 40 °C

i) The <u>heat change of the hot water</u> = mass x c x ΔT

mass = 50.0 g (you used 50.0 mL and water has a density of 1.00 g/mL)

c = 1.00 cal/g °C

ΔT = Final temperature - initial temperature = 40° – 62° = -22 °C

Q = 50.0 × 1.00 x (-22) = -1100 calories

ii) The <u>heat change of the surroundings</u> = heat change of water + heat change of calorimeter

a. The <u>heat change of the cold water</u> = mass × c × ΔT

mass = 50.0 g

c = 1.00 cal/g °C

ΔT = Final temperature - initial temperature = 40° – 22° = 18 °C

Q = 50.0 × 1.00 × 18 = 900 calories

b. The <u>heat change of the calorimeter</u> = (mass × c) × ΔT

ΔT is the same as for the cold water (18 °C), and the mass and c of the calorimeter will be constant throughout the day—the heat capacity.

Combining all of the above in equation 5:

$$-1100 = -[(900) + \{m(cal) \times c(cal)\} \times 18)$$

$$(m_{cal} \times c_{cal}) = 200 \div 18 = \mathbf{11.11} \text{ cal/°C}$$

This is the number (the heat capacity of the calorimeter) that should be recorded on line (*4*) of the data sheet and used for the rest of the lab.

B. Specific Heat of a Metal

In this experiment, the metal is our "reaction", and the water and calorimeter combine to give us our surroundings:

heat change of metal = - heat change of surroundings *(6)*

When I did this experiment, my data were:

5. Mass of metal = 30 g
6. Temperature of water before = 22 °C
7. Temperature of metal before = 97 °C
9. Maximum temperature = 32 °C (from a graph that is not shown!)

i) <u>the heat change of the metal</u> **= mass x c x ΔT**

= 30 g × c(metal) × (32°- 97 °C)
= 30 g × c(metal) × (-65 °C)
= -1950 × c(metal) calories

ii) <u>heat change of the surr.</u> = heat change of cold water + heat change of calorimeter

a. The heat change of the cold water = mass × c × ΔT

mass = 30.0 g (as the density of water is 1.00 g/mL)

c = 1.00 cal/g °C

ΔT = 32 - 22 °C = 10 °C

b. The heat change of the calorimeter = (heat capacity) × ΔT

ΔT is the same as for the cold water (26 °C), and the (heat capacity) of the calorimeter is as measured in Part A (11.11 cal/°C)

Combining all of the above in equation 6:

$$-1950 \times c(metal) = -[\{30 \times 1 \times 10\} + \{11.1 \times 10\}]$$

$$1950 \times c(metal) = (300 + 111)$$

$$c(metal) = 411 \div 1950$$

$$= \mathbf{0.21} \text{ cal/g °C}$$

C. Molar Heat of Solution

In this procedure, we can consider the process of dissolution as our reaction:

heat of solution = -heat change of surroundings *(7)*

When I did the experiment using NaOH:

11. Temperature of water before = 22 °C
12. Mass of solid = 1.402 g
13. Temperature of water after = 48 °C

i) <u>heat change of surroundings</u> = heat change of water + heat change of calorimeter

a. <u>heat change of water</u> $= m \times c \times \Delta T$
$= 25.0 \times 1 \times (48° - 22\ °C)$
$= 25 \times 26 = 650$ calories

b. <u>heat change of calorimeter</u> $=$ (heat capacity) $\times\ \Delta T$
$= 11.11 \times 26\ °C = 288.9$ calories

ii) <u>heat of solution</u> = -(650 + 288.9) = -938.9 calories

Thus, 1.402 grams of NaOH produces 938.9 calories when dissolved in water. This is the number that goes into line (*14*). However, we want the molar heat of solution - the heat given off when 1 mole of NaOH is dissolved in water. To get this value, we must convert to moles:

iii) Line 15. Molar mass of NaOH = 22.99 + 16 + 1.01 = 40 g/mole

Line 16. Number of moles of NaOH = g of NaOH ÷ molar mass of NaOH
= 1.402 g ÷ 40 g/mole = 0.0350 moles

The heat of solution is therefore -938.9 calories per 0.0350 moles of NaOH. To get the heat per mole, we simply divide by the number of moles:

Line 17. Molar heat of solution = -938.9 ÷ 0.0350 = -26,750 calories/mole
= **-26.750** kilocalories/mole

D. Energy of Reaction

In this case, we actually have a reaction, and the total water and calorimeter constitute our "surroundings":

heat change of reaction = - heat change of surroundings *(8)*

When I did the experiment, my results were:

18. Initial temperature = 22 °C
19. Final temperature = 52 °C

i) The heat change of the surroundings = heat change of water + heat change of calorimeter

a. The heat change of the water = mass × c × ΔT

mass = 102.0 g (the density of water is 1.00 g/mL and we have 50 + 52 mL of solution)

c = 1.00 cal/g °C

ΔT = 52° - 22 °C = 30 °C

b. The heat change of the calorimeter = (heat capacity) × ΔT

ΔT is the same as for the water (30 °C), and the heat capacity of the calorimeter is 11.11 cal/°C

Combining all of the above in equation 8:

heat change of reaction = - [{102 × 30} + {11.11 × 30}] = -3393 cal

As we want the heat PER MOLE, we must convert the above number, as follows:

Number of moles of HCl (the limiting reactant) = 1 M × 0.05 L = 0.05 moles

Thus, the molar heat of solution = our heat ÷ number of moles
= -3393 cal ÷ 0.05 moles
= -67,860 cal/mole = -67.860 kcal /mole

Light from Atoms

Introduction

Nearly everyone has seen the colors associated with certain elements during chemical reactions. These appear in a most spectacular fashion in the discharging of fireworks, but the same phenomenon can be observed in specially-treated fireplace logs which exhibit colors when burned and even in "neon" signs. The origin of these colors lies in the excitation of atoms by heat energy or electrical energy and the subsequent release of the excitation energy as photons of light. We can represent this process generally as shown below:

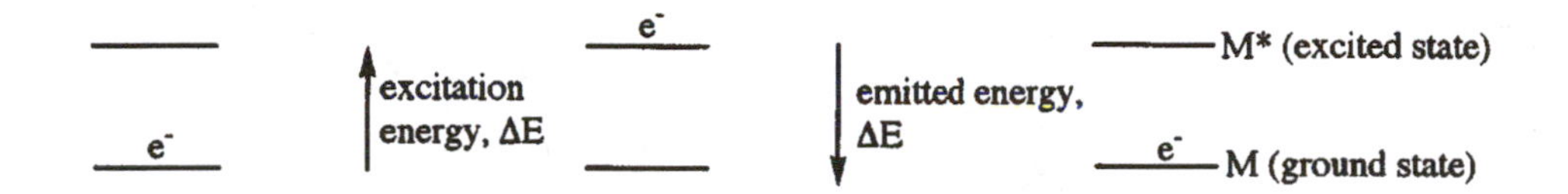

In the ground state, all of the electrons of atom M occupy orbitals with the lowest possible energy (according to the Aufbau principle), but when the atom absorbs enough energy, one or more of the valence electrons moves to a higher energy orbital. At this point, the atom is said to be in an excited state (M*). Such states are short-lived and rapidly decay back to the ground state by releasing energy in the form of light. The energy and frequency of the light which is released during the decay transition depend on the difference in energy (ΔE) between the ground state and the excited state.

For each atom which undergoes a transition from an excited state to a ground state, one photon is released with the transition energy, ΔE. Since every atom has many higher energy levels to which an electron may be excited, the light emitted from the various excited atoms will consist of several different energies or frequencies.

Planck's equation gives the relationship between the energy of a light photon (E) and its frequency (ν):

$$E_{photon} = h \times \nu \text{ (h = Planck's constant, } 6.63 \times 10^{-34} \text{ Js/photon)}$$

Spectroscopists often express the wave properties of a photon in terms of the wavelength (λ), rather than the frequency (ν). The two are related as:

$$c = \lambda \text{ (cm)} \times \nu(\text{sec}^{-1}) \quad [\text{c is the speed of light, } 3.0 \times 10^{10} \text{ cm/sec}]$$

Thus, the energy of a photon can be related to the wavelength by:

$$E = hc/\lambda$$

If a particular set of photons emitted from an excited element has a wavelength that falls within the visible spectrum (400–800 nm), our eyes will detect the light. The color we perceive will depend upon the wavelengths of the photons present. Each type of atom will have its own combination of wavelengths, and so we might expect to see a different color for each element. While our eyes do not discriminate among frequencies well enough for this to be entirely true, there are many elements that do exhibit characteristic colors when excited (for example, an intense yellow color from sodium). Such elements may be detected in samples by using a flame test—the sample is simply heated and the resultant color observed. In the first part of this experiment, you will observe the flame-test colors of several different elements.

A more detailed study of the light emitted from excited atoms can be done using a **spectroscope**—a device that separates the individual wavelengths present in light, usually via diffraction. A sketch of such a spectroscope is shown below. Light enters through a slit, is reflected from a diffraction grating and strikes a screen. Thus an **emission spectrum** is shown in the screen, consisting of colored lines for the different wavelengths. In the second part of this experiment, you will examine two forms of commercial light sources (fluorescent and incandescent) using a spectroscope.

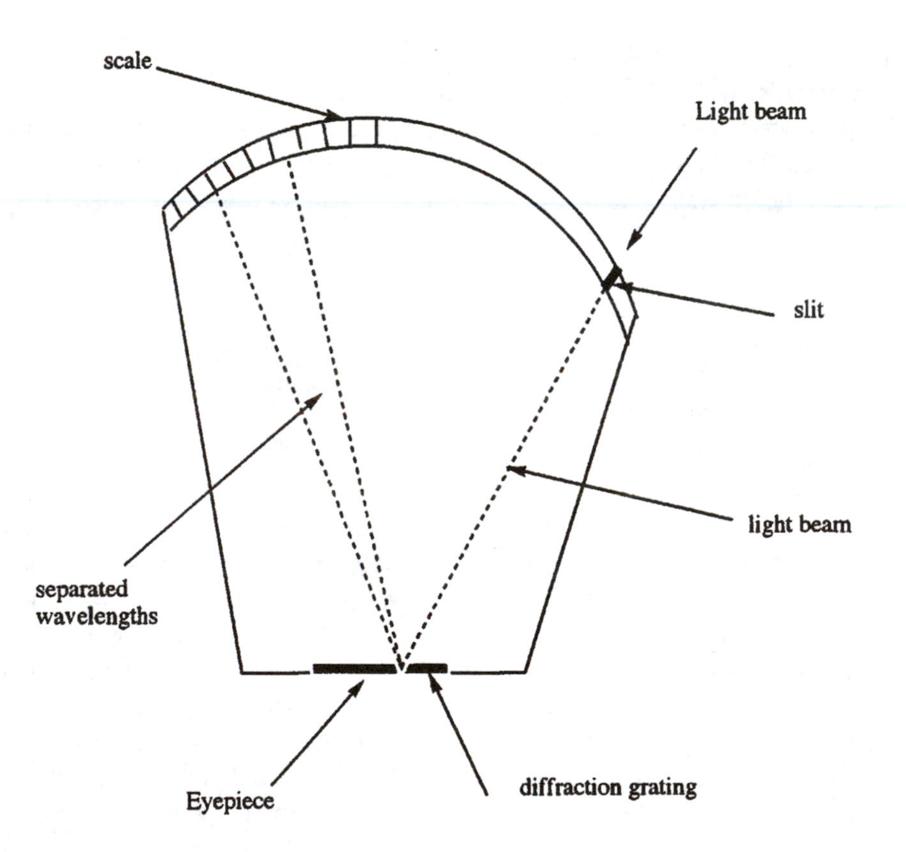

Early scientists attempted to relate the wavelengths (or frequencies) of different emission spectra to the atomic energy levels described in the Bohr model of the atom. While this did not work for most elements, a decent mathematical description of the hydrogen atom spectrum was generated:

$$\nu = \mathfrak{R}(1/n_1^2 - 1/n_2^2)$$

where $\mathfrak{R}$ is the Rydberg constant (3.25×10^{15} Hz) and n_1 and n_2 are the numbers of the appropriate energy levels (principle quantum numbers): n_2 is the higher energy level and n_1 is the lower energy level (thus, a photon is emitted as an electron drops from the n_2 level to the n_1 level).

In the final part of this experiment, you will examine the hydrogen spectrum using your spectroscope and attempt to assign quantum numbers to the excited and ground states of each line you observe. You will also have the opportunity to compare qualitatively the spectra of different elements.

PROCEDURE

A. *Flame Tests*

1. Your TA will perform flame tests on a number of metal chloride solutions [sodium, potassium, calcium, strontium and copper(II)]. Observe closely and record the color of each (*1*).

B. *Using the Spectroscope*

2. Obtain a spectroscope from the TA. THIS SHOULD BE HANDLED CAREFULLY AND KEPT DRY.
3. Point the slit towards a fluorescent ceiling light and look through the small circular opening of the grating. You should see light through the slit on the right and scales on the left as shown below. The light by the scales should consist of a continuous "rainbow" of color and three sharp lines. The lower scale is wavelength in nm.

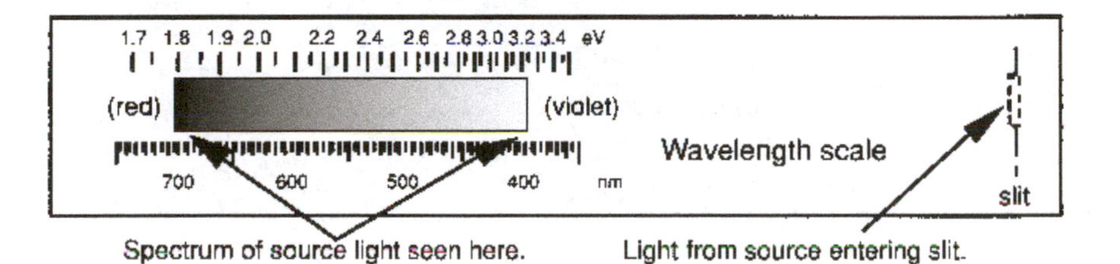

4. One of the lines you see should be a green one at 546 nm. If this is not the case, ask your TA to calibrate the spectroscope for you. When you are happy with the calibration, record the colors and wavelengths of the three lines (*2*).
5. Point the spectroscope towards the incandescent lamp set up in the lab. Describe/sketch the spectrum (*3*).
6. Place a piece of blue glass between the incandescent lamp and your spectroscope and describe/sketch the spectrum now (*4*).

C. *Atomic Spectra*

7. Your TA will set up a hydrogen discharge tube in the lab. DO NOT TOUCH THIS!!! View the tube from a distance of 1–2 meters through your spectroscope and record the wavelengths of all lines you observe (*5*). Using the fact that all the lines you see involve transitions from higher energy levels down to the n = 2 level, try to assign quantum numbers to each of the lines you observe (*5*).
8. Using the fact that the n = 2 level has an energy of -5.45×10^{-19} J per atom, construct a partial energy level diagram for the hydrogen atom (*6*). Label the transitions between levels according to your data in line *5*.

9. View the sodium flame test (conducted by your TA) through the spectroscope. Describe the spectrum and compare it to that of hydrogen (7).

SAMPLE CALCULATIONS

There is a very dim line in the hydrogen spectrum at 410 nm that is often hard to see. We shall use this one as the example.

Data Sheet Line 5:

$\lambda = 410 \text{ nm} = 410 \times 10^{-7} \text{ cm}$

$c = \lambda \text{ (cm)} \times \nu(\text{sec}^{-1})$

so $\nu = (3 \times 10^{10} \text{ cm/sec}) \div (410 \times 10^{-7} \text{ cm}) = 7.3 \times 10^{14} \text{ Hz (or sec}^{-1})$

As: $\nu = \Re(1/n_1^2 - 1/n_2^2)$ and we know that $n_1 = 2$

$7.3 \times 10^{14} = 3.25 \times 10^{15}(1/4 - 1/ n_2^2)$

$(1/4 - 1/ n_2^2) = 0.2246$

$1/ n_2^2 = 0.0254$

$n_2^2 = 40$

$\mathbf{n_2 \approx 6}$

These data have been filled in for you on the data sheet.

In order to find the energy of this transition (ΔE), use:

$E = h \times \nu$

$\Delta E = 6.63 \times 10^{-34} \times 7.3 \times 10^{14} \text{ sec}^{-1}$

$\Delta E = \mathbf{4.9 \times 10^{-19} \text{ J}}$

Thus, the energy difference between the second (n = 2) and sixth (n = 6) energy levels is 4.9×10^{-19} J.

In order to plot this sixth level on your energy level diagram (*6* on the data sheet):

We know that n = 2 has an energy of -5.45×10^{-19} J.

Thus, the n = 6 level has an energy of $(-5.45 \times 10^{-19} \text{ J}) + (4.9 \times 10^{-19} \text{ J}) = -0.6 \times 10^{-19}$ J.

This is the value that should be entered on your energy level diagram.

Periodicity and the Periodic Table

INTRODUCTION

By now, you have all been introduced to the Periodic Table of the elements. This Table is the cornerstone of chemistry, and its development is a perfect reflection of science in general, a discipline filled with lucky discoveries, brilliant work and controversy. In today's laboratory, you will perform a few experiments designed to demonstrate the form of the Table.

The Periodic Table is an arrangement of elements based on their properties and electronic structure. The electronic structure and many of the properties such as ionization energy and atomic radius are too complex to measure in a General Chemistry laboratory, so you will concentrate on physical properties and reactivity. Each stage of the lab requires that you observe particular properties and, from these observations, make generalizations about the trends of a property around the Table. Hopefully, by the end of the experiment, you will have a better feel for the entire concept of the Periodic Table.

Before you begin, you should note that we are going to ignore certain parts of the Table (the so-called transition metals, lanthanides and actinides) as well as elements discovered after 1870.

PROCEDURE

A. Physical Properties —To be Done Outside The Lab

Do @Home

1. The Houston Science Museum has a display of chemical elements. Alternatively, there are some nice visual Periodic Tables on the Web (for example, **periodic.lanl.gov**). At some stage before next week, go and look at one of these, and describe the appearance of the elements listed in the data sheet (*1*).
2. Using either the same source as in step 1, your textbook or some other resource, fill in the melting and boiling points of the elements listed in the data sheet (*2*) (ignore the X's for now!)
3. Examine both of your "tables" developed above and look for trends. Answer the questions on the data sheet (*3* through *5*).

B. Oxides and Hydroxides of the Elements

One of the original properties used to generate the periodic table was whether solutions of oxides of the elements in water were acidic or basic. In this part of the lab, you will "make" solutions of two oxides, and examine four others. You will use litmus paper to determine whether a solution is acidic or basic (red litmus turns blue in bases; blue litmus turns red in acids).

4. Take a small amount of magnesium oxide (made as in Experiment 3) and add it to 25 mL of water in a 100 mL beaker.
5. Add a small piece of dry ice (solid CO_2) to 25 mL of water in a different 100 mL beaker.

CAUTION! **DRY ICE CAN BURN SKIN—USE TONGS TO HANDLE**

6. Test these two solutions and the four laid out in the lab (solutions of oxides of Na, B, P, S—do not discard these) with litmus paper. Record your results (*6*).

C. *Reactivity of a Metal Group*

7. Half-fill three 150 mL beakers with water, cover with watch glasses and take to the hood.
8. Obtain a piece of Li (lithium) from your TA and add it to one beaker. Record your observations (*8*).
9. Obtain a piece of Na (sodium) from your TA and add it to a second beaker. Record your observations (*9*).

CAUTION! **Na and K REACT VIGOROUSLY WITH WATER**

10. Obtain a piece of K (potassium) from your TA and add it to the third beaker. Record your observations (*10*).

D. *Reactivity across a Period*

11. Half-fill four 150 mL beakers with water and cover with watch glasses.
12. Place a small piece of Mg, Al, Si and S into a beaker each, replace the watch glass and observe and record (*12*). Wash and "recycle" any remaining silicon.
13. Put about 5 mL of a 6 M HCl solution into each of four beakers.
14. Place a small piece of Mg, Al, Si and S into a beaker each, cover with a watch glass and observe and record (*12*). Wash and "recycle" any remaining silicon.
15. There is a sample of phosphorus laid out for you to examine. Do not touch it, but infer information from the sample to fill in the appropriate blank in line (*12*).

E. *Reactivity of a Non-Metal Group*

In this part of the experiment, you will examine the relative reactivity of chlorine, bromine and iodine (halogens). In order to do this, you will combine each element with another which is in the form of its anion. If the elemental halogen (Halogen 1) is more reactive than the anionic halogen (Halogen 2), a reaction will occur; Halogen 1 will form its anion, and Halogen 2 will revert to the elemental state. Thus, for example, if F (fluorine) is more reactive than chlorine (Cl), the following would be the case:

$$2\ Cl^-(aq) + F_2(aq) \rightarrow Cl_2(g) + 2\ F^-(aq)\ \text{(Reaction)}$$

$$2\ F^-(aq) + Cl_2(aq) \rightarrow \text{No Reaction}$$

You will make and use aqueous solutions of the three halogens. (Does this make sense in light of your observations in the previous part of the lab?)

THE ELEMENTAL HALOGENS ARE TOXIC. DO NOT BREATHE OR HANDLE!

16. To Make Chlorine Water: **IN THE HOOD**, add 50 mL of water, 20 mL of bleach and 15 mL of 2 M HCl to a 250 mL beaker. Cover with a watchglass and swirl. (THIS MAY HAVE BEEN DONE FOR YOU).
17. To Make Bromine Water: Add 150 mL of 0.2 M KBr (potassium bromide) solution, 0.1 g of $KBrO_3$ (potassium bromate) and 25 mL of 2 M HCl to a 400 mL beaker. Cover with a watchglass and swirl. (THIS MAY HAVE BEEN DONE FOR YOU).
18. To Make Iodine Water: Add 150 mL of water, 5 mL 0.2 M KI (potassium iodide) solution, 0.1 g of KIO_3 (potassium iodate) and 25 mL of 2 M HCl to a 400 mL beaker. Cover with a watchglass and swirl. (THIS MAY HAVE BEEN DONE FOR YOU).
19. Record the appearance of each solution (*14*).
20. Take 2 mL of each solution in different test tubes. Add 2 mL of cyclohexane to each and record the color of the solution (*14*).

DISPOSE OF CYCLOHEXANE IN ORGANIC WASTE BOTTLE!

21. Add 5 mL 0.2 M KBr solution to a test tube. Add 5 mL of chlorine water, stopper and shake. Record your observations (in particular, color changes) (*14*).
22. Add 5 mL cyclohexane to the tube. Record the color change (*14*).
23. Add 5 mL 0.2 M KI solution to a test tube. Add 5 mL of chlorine water, stopper and shake. Record your observations (in particular, color changes) (*14*).
24. Add 5 mL cyclohexane to the tube. Record the color change (*14*).
25. Add 5 mL 0.2 M KCl solution to a test tube. Add 5 mL of bromine water, stopper and shake. Record your observations (in particular, color changes) (*14*).
26. Add 5 mL cyclohexane to the tube. Record the color change (*14*).
27. Add 5 mL 0.2 M KI solution to a test tube. Add 5 mL of bromine water, stopper and shake. Record your observations (in particular, color changes) (*14*).
28. Add 5 mL cyclohexane to the tube. Record the color change (*14*).
29. Add 5 mL 0.2 M KCl solution to a test tube. Add 5 mL of iodine water, stopper and shake. Record your observations (in particular, color changes) (*14*).

30. Add 5 mL cyclohexane to the tube. Record the color change (*14*).
31. Add 5 mL 0.2 M KBr solution to a test tube. Add 5 mL of iodine water, stopper and shake. Record your observations (in particular, color changes) (*14*).
32. Add 5 mL cyclohexane to the tube. Record the color change (*14*).

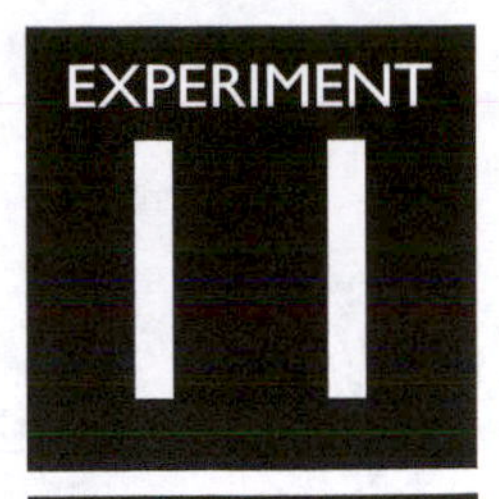

Properties Associated with Changes in Physical State

INTRODUCTION

From your own experience, you know that substances at ordinary temperatures and pressures exist in one of three physical states or phases: solid, liquid, or gas, and that changing the temperature or pressure of a substance can often cause it to change from one phase to another.

The change associated with the transformation from the solid to the liquid state is called **melting** or **fusion,** and the reverse transformation from liquid to solid is called **freezing.** When a solid is heated, its temperature rises until some of it begins to melt. If it is a pure substance, the temperature remains constant as more heat is added until all of the solid has liquified. The temperature at which this occurs is called the **melting temperature** of the substance. Similarly, when a liquid is cooled, its temperature decreases until some of it freezes. If the substance is pure, the temperature remains constant as more heat is removed until all of the liquid has solidified. The temperature at which this occurs is called the **freezing temperature** of the substance. For all common substances, the melting temperature at a given pressure is the same as the freezing temperature. This temperature is only slightly pressure dependent. Melting and freezing temperatures measured at 1 atmosphere pressure are called **normal melting** or **freezing points.**

The change associated with the transformation from the liquid to the gaseous state is called **vaporization** and the reverse transformation from gas to liquid is called **condensation.** As with freezing and melting, the temperature of a pure substance remains constant while vaporization or condensation is taking place and this temperature is called the **boiling temperature.** The boiling temperature is much more pressure dependent than the melting temperature. When measured at 1 atmosphere pressure, the boiling temperature is called the **normal boiling point.**

The change associated with the transformation from the solid directly to the gaseous state (without melting) is called **sublimation**, and the reverse process is called **deposition.**

The melting and boiling points of pure substances are intensive physical properties. Since different pure substances usually have significantly different melting and boiling temperatures at a given pressure, values for these are frequently included in the collection of properties used to distinguish one substance from another. You already made use of this idea in Experiment 2. In today's experiment you will determine and compare the normal melting and boiling points of several substances.

PROCEDURE

A. Melting Point of Naphthalene

1. Fill a 150 mL beaker three quarters full of water to be used as a heating bath.

2. Obtain a few capillary melting point tubes and a small sample of naphthalene (about the size of a pencil head eraser) on a watch glass (use your scoopula to accomplish this task).

NAPHTHALENE IS TOXIC BY INGESTION AND SKIN CONTACT. BE CAREFUL!

3. Introduce some of the naphthalene into a capillary tube by pressing the open end of the tube into a small pile of the solid. Hold the tube vertically and gently tap the sealed end against the bench top until the solid is packed at the bottom of the tube. You should continue to do this until you have about 5 mm of solid in the tube.
4. Attach the filled tube to a thermometer using a rubber band. The thermometer and the capillary tube are fragile—HANDLE THEM GENTLY. The solid should be next to the thermometer bulb and the open end of the tube should face upwards as shown below:

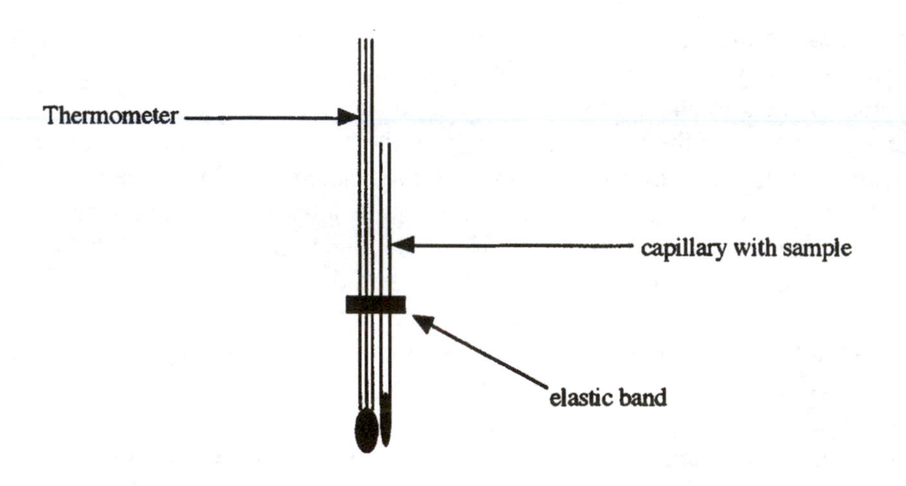

5. Clamp the thermometer and attached tube in the water bath. The assembly should be suspended in the water bath so that it is touching neither the side nor the bottom of the beaker. Make sure that the top of the tube remains well above the water line while the bottom is submerged.
6. Heat the bath GENTLY with a Bunsen burner flame (about 5° a minute), stirring as you do. Watch the solid carefully.
7. As the bath temperature approaches 80 °C (the anticipated melting point), slow the heating process. When the solid begins to turn into a liquid (not just softens) record the temperature (*1*).
8. Dispose of the capillary tube in the WASTE GLASS container in the lab and the naphthalene in the SOLID WASTE bottle in a hood.
9. Turn the burner off and allow the bath to cool to less than 50 °C before measuring the sample in Section B.

B. Boiling Point of Methanol

10. You will use the same form of apparatus to measure boiling points as you used to determine the melting point. The one difference is to use a small culture tube instead of the capillary tube, which you should attach with an elastic band to the thermometer.

11. Obtain about one mL of methanol in a small beaker.

METHANOL IS FLAMMABLE AND SHOULD BE CONSIDERED TOXIC BY INHALATION, INGESTION, OR SKIN CONTACT!

12. Pipet a few drops of methanol into the culture tube. Place a capillary tube into the culture tube so that the open end is immersed in the liquid:

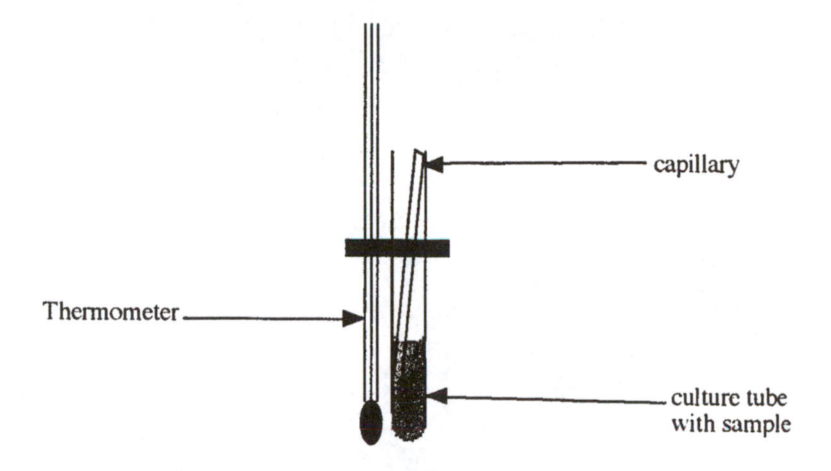

13. Position the thermometer and culture tube in the water bath. DO NOT LET WATER ENTER THE CULTURE TUBE!
14. Heat the water bath slowly using a Bunsen burner. You should see bubbles of trapped air start to come out of the capillary tube. Stir the water bath until you observe a rapid and continuous stream of these bubbles.
15. Stop heating the water bath and stir as it cools. Watch the capillary tube closely. Note the temperature at which the bubbles stop and methanol begins to enter the capillary tube. This is the boiling temperature of the methanol (*2*). This value should be about 65 °C.
16. Turn the burner off and allow the bath to cool to less than 50 °C before measuring the sample in Part C. Dispose of any residual methanol in the LIQUID WASTE BOTTLE.

C. Boiling Point of an Unknown Liquid

17. Obtain about 1 mL of a liquid "unknown" as assigned by your TA. Record its code number in line *3*.

ASSUME THAT THE UNKNOWN IS FLAMMABLE AND SHOULD BE CONSIDERED TOXIC BY INHALATION, INGESTION, OR SKIN CONTACT.

18. Repeat steps 12–16, recording the observed boiling temperature in line *4* of the data sheet.
19. Dispose of any residual liquid unknown in the LIQUID SOLVENT BOTTLE. Return the WASHED culture tube to the place from which you obtained it.

Determination of Molar Mass by Freezing Point Depression

INTRODUCTION

Solutions are homogeneous mixtures of two or more pure substances. The major component of a solution is called the **solvent** and the minor components are called **solutes.** When a solution is composed mostly of solvent (dilute solution), the physical properties of the solution may resemble, but are not exactly the same as, those of the pure solvent. Some of the physical properties of a dilute solution are independent of the type of solute and depend only on the number of solute particles present in the solution. Such physical properties are called **colligative properties.** Vapor pressure, boiling point and freezing point are examples of colligative physical properties. Experiments have shown that, when a nonvolatile solute is dissolved in a solvent, the vapor pressure of the solution is lower than that of the pure solvent. This in turn lowers the freezing point and raises the boiling point of the solvent. The extent of these changes is dependent on how much solute is present (the concentration of the solute). In this experiment, you will investigate the lowering of freezing point.

For a nonvolatile, non-electrolyte solute, the change in freezing point (ΔT_f) can be represented by equation 1:

$$\Delta T_f = (K_f)(m) \qquad (1)$$

where m is the molality of the solution (moles of solute/kilograms of solvent) and K_f is a constant specific for each solvent. Using the definition of molality and the fact that moles of a given substance = (grams of substance/molar mass of substance), equation 1 can be rewritten:

$$\Delta T_f = [(K_f)(\text{grams of solute})] \div [(\text{molar mass of solute})(\text{kilograms of solvent})] \qquad (2)$$

Equation 2 can be rearranged and written in terms of the molar mass of the solute:

$$(\text{molar mass of solute}) = [(K_f)(\text{grams of solute})] \div [(\Delta T_f)(\text{kilograms of solvent})] \qquad (3)$$

Thus, if one takes a known mass of a solute dissolved in a known mass of a solvent (for which the value of K_f is known) and measures the change in freezing point between the solvent and solution, substitution of these values into equation 3 will lead to the molar mass of the solute.

In this experiment, you will examine all aspects of this method. In the first part, you will determine the freezing point of a pure solvent, cyclohexane. In the second part, you will determine the value of K_f for cyclohexane by measuring the freezing point of a solution made by adding a known amount of a known compound to the cyclohexane. Finally, in the third part of the experiment, you will make a solution of an unknown solid in cyclohexane, determine the freezing point of the solution, and thereby determine the molar mass of the unknown.

PROCEDURE

A. *Freezing Point of Cyclohexane*

1. Nearly fill a 400 mL beaker with ice and water.
2. Place a test tube in a 250 mL beaker and record the mass (*1*).
3. Add about 20 mL of cyclohexane to the test tube and reweigh (*2*).

CYCLOHEXANE IS FLAMMABLE AND TOXIC!

4. Immerse the test tube vertically in the ice bath and clamp.
5. While stirring the cyclohexane, record the temperature every 30 seconds (*3*). You should reach a point at which the temperature holds constant as the solvent freezes. When it starts to dip again, stop recording and remove the test tube from the ice bath.
6. Plot your results with time on the x-axis and temperature on the y-axis (see the sample calculations section). Record your freezing point (*4*).

B. *Freezing Point of a Naphthalene Solution*

7. Weigh out about 0.2 g of naphthalene on weighing paper and record the exact mass (*5*). Pour it **all** into the test tube containing cyclohexane from the first part (which you have allowed to melt) and stir until completely dissolved.

NAPHTHALENE IS TOXIC BY INGESTION AND SKIN CONTACT!

8. Perform steps 4–6 with this solution, recording your results on the data sheet (*6a*).
9. Allow all of the cyclohexane to melt.
10. Weigh out another portion of the naphthalene of about 0.2 g, record (*5*) the TOTAL mass (*i. e.,* this amount plus that from the first try), and add to the tube.
11. Perform steps 4–6 with this new solution. Record in *6b* on data sheet.
12. Plot the data from steps 8 and 11 on the same graph as step 6. Determine the freezing points (*7*).
13. Calculate the freezing point depression constant of cyclohexane (*8*). Calculate your average value (*9*).
14. Dispose of cyclohexane in waste bottle.

C. *Freezing Point of an Unknown Solution*

15. Repeat steps 2 and 3, recording the values in lines *10* and *11*, respectively on the data sheet.
16. Repeat steps 7 through 12 using an unknown solid [record code number (*12*)] instead of the naphthalene. Record the mass of the unknown in line *13*, the temperatures in lines *14a* and *14b*, and put the freezing points in line *15*. If necessary, use a new piece of graph paper!

17. Determine the molar mass of the unknown (*16*). Give this value to your TA. He/she will post all results outside the lab. Return during the week, copy the ones for the unknown that you used (*17*) and calculate the mean (*18*) and standard deviation (*19*) for all results of your unknown.
18. Dispose of cyclohexane in waste bottle.

Sample Calculations

Graphing Data

A typical graph for one trial "should" look like:

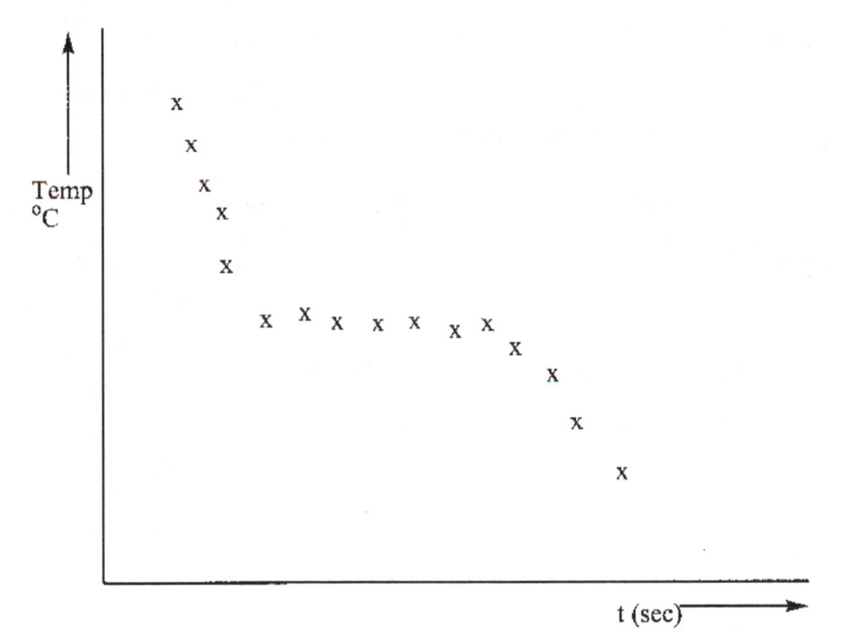

In order to obtain the freezing point, draw straight lines through both the initial slope of the data and the "flat bit" as shown:

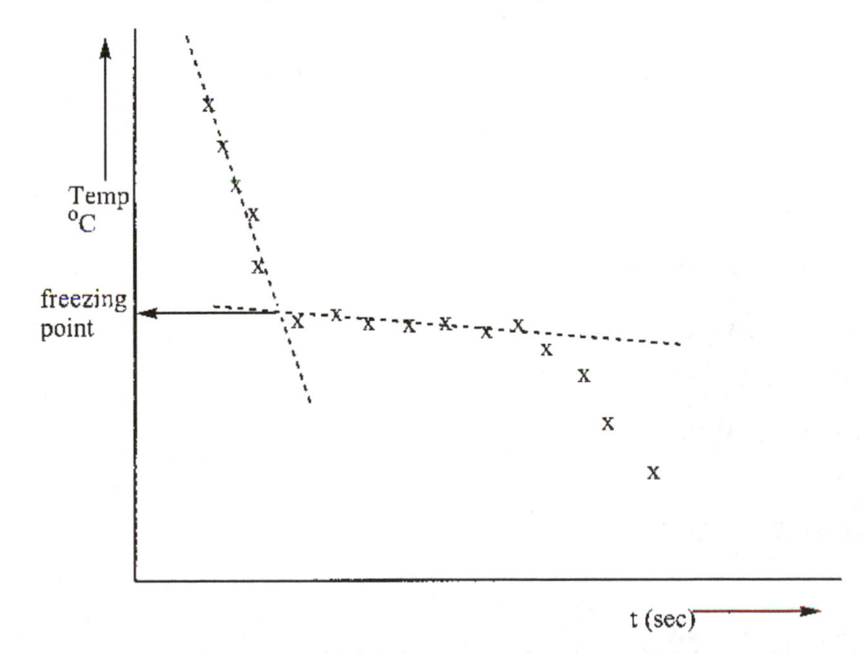

The intersection of the lines corresponds to the freezing point—marked on the second diagram.

When I did a similar experiment using a different solvent:

A. Freezing point of Solvent

Return Large test tube to #300!

1. Mass of test tube and beaker 20.3 g
2. Mass of test tube, beaker and solvent 20.9 g
4. Freezing point of pure solvent (°C) 52 °C (from a graph)

B. Freezing point of Naphthalene Solution

5. Mass of naphthalene 0.1 g
7. Freezing point 50 °C (from a graph)

To determine the K_f of the solvent, use equation 2:

1*

ΔT_f = [(K_f)(grams of solute)] ÷ [(molar mass of solute)(kilograms of solvent)]

ΔT_f = 52 ° – 50 ° = 2 °C

grams of solute = 0.1 g

molar mass of solute ($C_{10}H_8$) = (12.01 × 10) + (1.01 × 8) = 128.18

kilograms of solvent = 0.0006 kg [(*2* – *1*) / 1000]

Thus:

2 = [(K_f)(0.1)] ÷ [(128.18)(0.0006)]

K_f = 1.54 °C/m

(this number goes in line *8*, and your average value in line *9*)

C. Freezing Point of an Unknown Solution

10. Mass of test tube and beaker 20.6 g
11. Mass of test tube, beaker and cyclohexane 21.3 g
13. Mass of unknown 0.1 g
15. Freezing point 48.5 °C (from a graph)

To determine molar mass, use equation 3:

2*

(molar mass of solute) = [(K_f)(grams of solute)] ÷ [(ΔT_f)(kilograms of solvent)]

K_f = 1.54 °C/m (from above)

grams of solute (unknown) = 0.1 g

ΔT_f = 52 – 48.5 = 3.5 °C

kilograms of solvent = 0.7 g = 0.0007 kg [(*12* – *11*) / 1000]

Thus:

Molar mass of solute = [(1.54 °C/m)(0.1 g)] ÷ [(3.5 °C)(0.0007 kg)] = **62.9 g/mol**

Organic Chemistry

INTRODUCTION

Organic chemistry is the study of compounds that contain at least one carbon atom bonded to hydrogen. The field is immense, due both to the ability of carbon to bond to itself and to the fact that all living things are essentially "organic." There is obviously not enough time in one lab period for you to experience all of organic chemistry, so we will concentrate on some "real-life" examples—soap, nylon, and perfumes.

A. Soap

A "soap" is an example of an "amphiphilic" compound—it must have an affinity for both water and organic substances. A usual soap molecule, therefore, has a long "organic chain" ending with an ionic group (example shown below). The idea is that the long organic chain dissolves in the water-insoluble dirt or grease, presenting the ionic "head-group" to the outside. This ionic group is "mixable" with water, thus one has effectively made the dirt or grease, water-soluble.

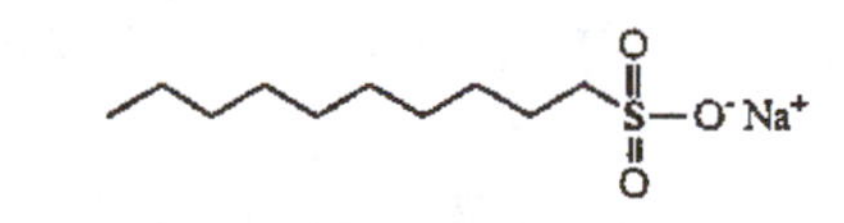

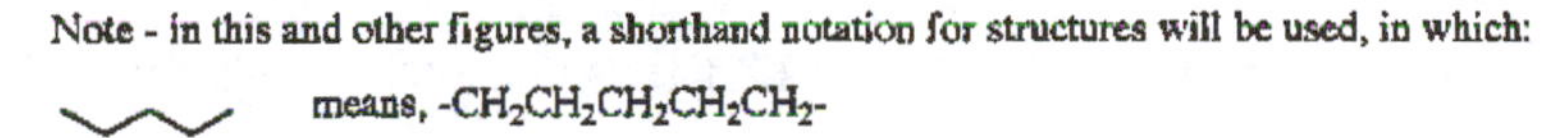

B. The Manufacture of Nylon

Nylon is a man-made polymer. A polymer is simply a large compound made up of repeating units, *e. g.*,

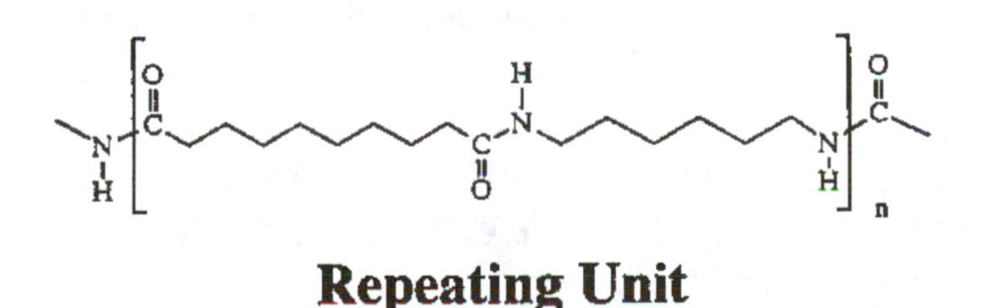

Repeating Unit

This is one example of nylon—Nylon 6-10 (as there are 6 and 10 carbon atoms in each repeating unit or monomer.) We make Nylon 6–10 according to the following condensation reaction:

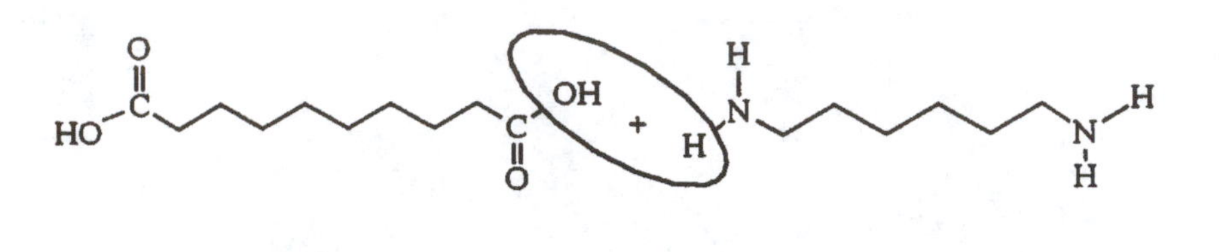

C. Perfumes

Many organic compounds have readily identifiable smells. For example, many "amines" (compounds containing -NH_2 groups) smell like rotten fish. Fortunately, however, many of the smells of organic compounds are pleasant and they can be used as perfumes—not just for cosmetic purposes, but also to add scent to numerous other household articles. In this part of the experiment, you will smell a variety of different compounds, and also make your own perfume.

Procedure

A. Soap

1. Place 100 mL beaker with 3 g of NaOH pellets, 5 g of lard and 5 mL water into a bigger, 400 mL beaker, half-filled with water.

SOLID NaOH IS CORROSIVE. DO NOT TOUCH THE PELLETS. RINSE THOROUGHLY WITH WATER IF YOU DO.

2. Heat with stirring for half an hour. Begin Part B, while you are waiting. Continue to check on your heating.
3. Transfer solution from smaller beaker into bigger beaker from step 1. Add 100 mL of NaCl solution and filter the soap. Allow to dry and examine it. Record your observations in data sheet (*1*).

B. Nylon

4. Obtain 12 mL of a solution of hexamethylenediamine in water in a 100 mL beaker.

HEXAMETHYLENEDIAMINE IS A STRONG IRRITANT AND POISON. IF YOU SPILL IT ON YOUR SKIN, WASH IT OFF WITH LOTS OF WATER.

5. Add 1 g of sodium carbonate and stir until it has dissolved.

6. In a separate beaker, obtain 25 mL of a solution of sebacoyl chloride in cyclohexane.

SEBACOYL CHLORIDE IS POISONOUS AND CAUSES SEVERE BURNS. SKIN CONTACT MAY PROVE FATAL!! CYCLOHEXANE IS FLAMMABLE AND TOXIC. IF YOU SPILL ANY OF THESE TWO CHEMICALS, WASH WITH LOTS OF WATER.

7. Slowly and carefully pour the cyclohexane solution into the beaker containing the aqueous solution, trying not to mix the two solutions. Note that a gummy solid starts to form at the interface between the two solutions.

8. Set up two stands with clamps and greased glass rods to act as a "pulley" system as shown below:

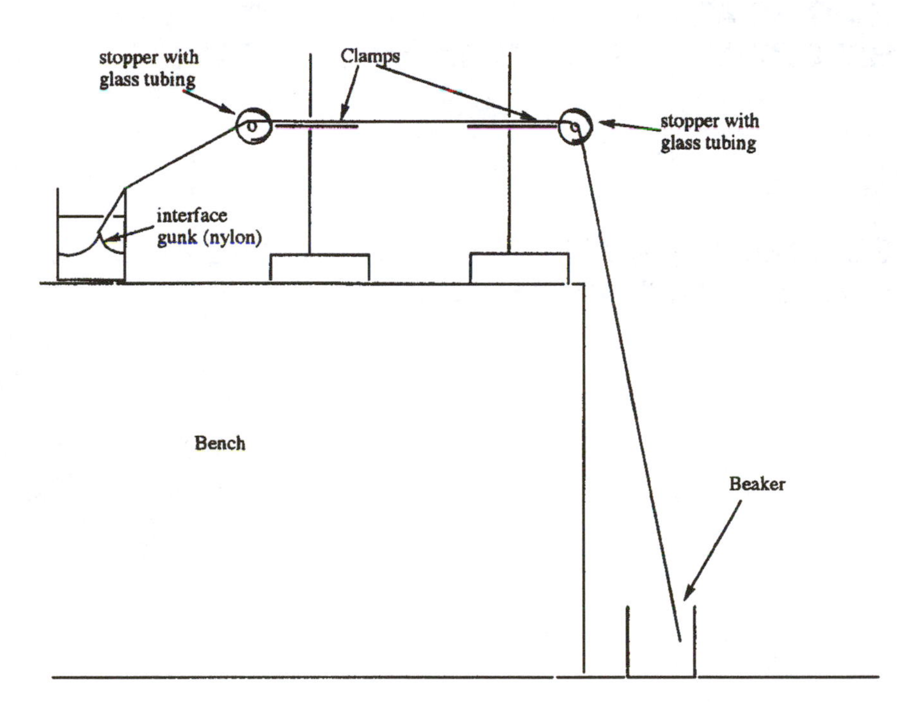

9. Bend a hook at the end of a piece of wire, and "snag" the interface film. Carefully pull out and drape the resulting "rope" over the two glass rods and into a beaker placed on the floor. The rope should now run from the interface into the beaker on the floor. If the rope breaks at any point, simply repeat this step.

10. Wash the rope THOROUGHLY with water, allow it to dry and then examine it. Record your observations on the data sheet (*2*).

C. Perfumes

11. Smell the samples of vanillin, methyl benzoate, benzaldehyde, methyl butyrate and cinnamic aldehyde laid out. Describe the odors (*3*).

REMEMBER TO SMELL IN THE CORRECT FASHION. WAVE YOUR HAND OVER THE TOP OF THE CONTAINER WHICH SHOULD BE HELD AT NOSE LEVEL ABOUT SIX INCHES IN FRONT OF YOUR FACE.

12. Add one drop of 3 M H_2SO_4, 2 drops of glacial acetic acid and 3 drops of 1-pentanol to a small test tube. Heat this in a water bath for 10 minutes and smell cautiously. Attempt to identify the product (*4*).

H_2SO_4 IS A VERY STRONG ACID DEHYDRATING AGENT. REPORT ANY SPILLS TO TA IMMEDIATELY!

GLACIAL ACETIC IS A VERY CORROSIVE ACID. REPORT ANY SPILLS TO TA IMMEDIATELY!

Kinetic Properties of a Chemical Reaction

INTRODUCTION

Beginning chemistry students sometimes have the mistaken idea that all chemical reactions happen instantly. This is probably because they have seen reactions such as fire or the precipitation of silver chloride, which appear to be almost instantaneous. However, one need only think of the rusting of an iron nail or the fermenting of fruit juice in wine making, which are very slow reactions, to realize that this impression is incorrect. As the speed or rate of reactions differ so greatly, therefore, the study of chemical rates or kinetics is a very important discipline. In this experiment, you will examine various aspects of the kinetics of a so-called "clock reaction":

$$S_2O_8^{2-}(aq) + 3\ I^-\ (aq) \rightarrow 2\ SO_4^{2-}\ (aq) + I_3^-(aq)$$

colorless colorless colorless brown

In this experiment, you will investigate the effects upon the rate of three different factors:

a) concentration of reactants;

b) temperature; and

c) the presence of a catalyst.

In order to accomplish this, you will perform the reaction four times, under different conditions, monitoring the progress of the reaction by observing the rate at which the brown triiodide ion appears. The four different sets of conditions are:

i) at 30 °C;

ii) at 30° C with reduced amount of $S_2O_8^{2-}$;

iii) at 37 °C; and

iv) at 37 °C with a catalyst present.

Thus, you will be able to observe the effect of changing concentration by comparing the results from trials (i) and (ii); the effect of temperature by comparing trials (i) and (iii); and the effect of a catalyst by comparing runs (iii) and (iv).

As this is a chemistry class, it is not sufficient just to observe the various effects qualitatively; we will want to express the results numerically. We shall accomplish this by measuring the concentration of the triiodide anion as time progresses, and converting these data to useable numbers. In order to measure the iodide ion concentration, we shall use the fact that the ion is colored—thus, the higher the concentration of iodide in a particular solution, the darker the color. Things appear colored because they absorb particular wavelengths of visible light and allow other wavelengths to pass through (to be transmitted). More intensely colored things absorb more light (have a higher absorbance, A). For solutions, the absorbance is directly related to the concentration, as described by Beer's Law (equation 1):

$$A = \varepsilon lc \tag{1}$$

(ε and l are constants associated with the particular substance and the distance over which the light passes through the substance, respectively).

Thus, the concentration of a colored substance may be determined by measuring the absorbance of the solution. We will use an instrument called a spectrophotometer (see diagram below) in order to measure the absorbance due to triiodide.

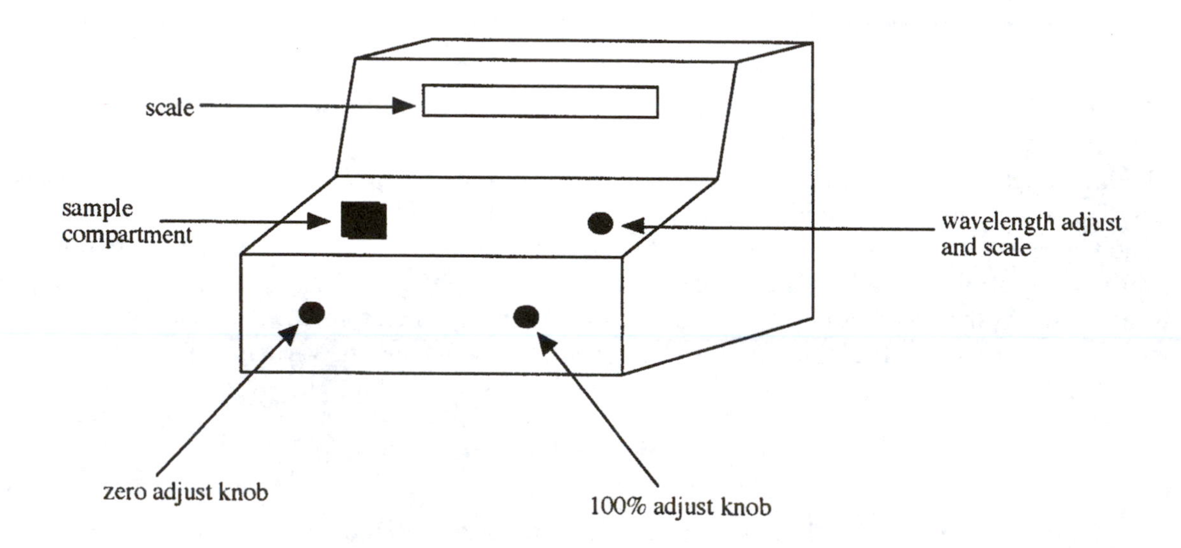

For technical reasons, a spectrophotometer uses a logarithmic scale to depict absorbance which is particularly hard to read accurately. Accordingly, therefore, you will actually measure the % transmittance of light (or how much light passes through the sample), %T. The scale for this on the spectrophotometer is linear and thus easier to read. The %T is related to the absorbance (A) according to:

$$A = \log (100 \div \%T) \tag{2}$$

Thus, each time you do the reaction, you will measure the % transmittance of your solution every minute or so, convert these values to absorbance using equation 2, and convert absorbance to concentration using equation 1. As you will use the same substance and test tube for each measurement, ε and l will not change during the course of the experiment. You will determine the value of εl by determining the absorbance of a solution for which you know the concentration of the triiodide ion (one in which the reaction has gone to completion).

Once you have a set of data that relate concentration and time, you will need to plot these on a graph. You will then use these graphs to determine the rate law and rate constant for each set of data. This process will be explained further in the section dealing with sample calculations. You will need to be familiar with rate laws and the Arrhenius equation, however.

PROCEDURE

You will do four "runs" during the course of today's experiment, each time varying certain conditions. The steps to be followed in each run are:

1. Take specified amounts (see Table) of the two solutions (0.020 M $(NH_4)_2S_2O_8$ and 0.30 M KI) in SEPARATE test tubes and place these in a water bath that is maintained at a particular temperature (see Table for temperatures).
2. While these solutions are heating, make sure that a spectrophotometer that is close to the water bath is set to 525 nm. Then use a third test tube (your "cuvette") to calibrate the spectrophotometer. This is a two step process.
3. Close the cover of the sample compartment and use the LEFT HAND KNOB to adjust the transmittance (the top scale) to zero.
4. Etch or draw a 1 cm vertical line on your cuvette starting at the top. This line will serve as your "placement marker."
5. Fill your cuvette two-thirds full with deionized water and place it in the sample compartment so that your placement marker is aligned with the mark on the compartment. Use the RIGHT HAND KNOB to adjust the transmittance to 100%.
6. When you have done this, put your cuvette in the water bath with the other two test tubes.
7. When the solutions in the test tubes have reached the temperature of the water bath, remove and dry the cuvette and pour both solutions into it. Quickly, measure and record the transmittance of the solution. Also measure and record the temperature of the water bath.
8. Replace the cuvette in the water bath for about 30 seconds and then remove it, dry it and take another measurement of the transmittance. Put the cuvette back in the water bath.
9. Repeat step 8 about 15 times (over the course of about 15 minutes).
10. When done, pour the solution from the cuvette into a new clean, labeled test tube and put this in the water bath. Clean and dry your cuvette and prepare for the next run.
11. Convert the transmittance to absorbance then to concentration of I_3^-. Plot a graph of $[I_3^-]$ against time, in minutes.

Run	Contents of first test tube	Contents of second test tube	Water bath
1	3 mL $(NH_4)_2S_2O_8$	3 mL KI	30 °C
2	1.5 mL $(NH_4)_2S_2O_8$+ 1.5 mL water	3 mL KI	30 °C
3	3 mL $(NH_4)_2S_2O_8$	3 mL KI	37 °C
4	3 mL $(NH_4)_2S_2O_8$	3 mL KI + drop of $Cu(NO_3)_2$	37 °C

Notes

12. In order to relieve congestion, half of you will do the runs in order 1, 2, 3, 4. The other half will do them in the order 3, 4, 1, 2. Be sure to use the correct parts of the data sheet.
13. When you are finished with ALL the runs, go back to the spectrophotometer which you used for your first run and calibrate it.
14. Take the solution left in step 10 from your FIRST run, pour it into your cuvette and measure its transmittance. Record this in line *5* of the data sheet.
15. Clean out all glassware thoroughly.
16. There are four areas on the data sheet in which to record your time and transmittance data. These areas also include spaces for you to write the absorbance and concentration information. You cannot do this until you have finished the entire experiment, however, and used the εl value calculated from the data for the "completed reaction."

SAMPLE CALCULATIONS

When I did the experiment using 3 mL of 0.035 M $(NH_4)_2S_2O_8$ and 3 mL of 0.3 M KI, my data for run 1 were:

Temperature = 30° throughout.			
Time (min)	Transmittance(%)	Time (min)	Transmittance (%)
1	73.4	11	48.1
2	66.9	12	47.6
3	64.8	13	47.1
4	59.2	14	46.4
5	55.5	15	46.7
6	54.9	16	46.4
7	53.9	17	46.4
8	52.3	18	46.4
9	50.0	19	46.4
10	49.0	20	45.1

After letting this reaction sit for 2 hours (so that it went to completion):

5. Transmittance of solution = 42%

Calculations

The first thing one has to do for all your transmittance data is convert to absorbance using equation 2. Thus, for the "completed reaction" (for which the transmittance was 42% above):

(See Intro.) 2 ☆ A = log (100 ÷ %T) = **0.377**. (record your value for this in line *6* of the data sheet)

We can also do this calculation for the transmittance data in the table above.

Having obtained absorbance data, we now need to convert it to concentration data, using equation 1. In order to do this, we need to know the value of the constant, εl.

We obtain the value of εl using the information for the run that went to completion, for which the absorbance is 0.377. By the time this measurement was made, we have to assume that all of the $S_2O_8^{2-}$ (the limiting reagent) was converted to I_3^-. In other words, **$[I_3^-]$ = initial $[S_2O_8^{2-}]$.**

To get the value of the initial $[S_2O_8^{2-}]$, remember we started with 3 mL of 0.035 M $(NH_4)_2S_2O_8$ and diluted this to 6 mL. Thus,

$[S_2O_8^{2-}]$ = (0.035 M) × (3 mL ÷ 6 mL) = **0.0175 M** (your value will be different, record in line *7*)

Using our absorbance data and substituting this into equation 1, we get:

$$A = \varepsilon lc, \text{ or } \quad 0.377 = \varepsilon l \times 0.0175 \text{ M}$$

$$\varepsilon l = \mathbf{21.5}$$ (record your value in line *8*)

We can now use this value to convert all of our absorbance data into concentration data. The completed version of my data table is shown below:

Time (min)	Transmittance	Absorbance	$[I_3^-]$
1	73.4	0.134	0.00625
2	66.9	0.175	0.00813
3	64.8	0.188	0.00875
4	59.2	0.228	0.0106
5	55.5	0.256	0.0119
6	54.9	0.260	0.0121
7	53.9	0.269	0.0125
8	52.3	0.282	0.0131
9	50.0	0.301	0.0140
10	49.0	0.310	0.0144
11	48.1	0.318	0.0148
12	47.6	0.323	0.0150
13	47.1	0.327	0.0152
14	46.4	0.333	0.0155
15	46.7	0.331	0.0154
16	46.4	0.333	0.0155
17	46.4	0.333	0.0155
18	46.4	0.333	0.0155
19	46.4	0.333	0.0155
20	45.1	0.346	0.0161

We now plot a graph of $[I_3^-]$ against time. Plot the points, then draw a smooth curve through the graph and *extrapolate this curve to time zero.* My plot is shown below (the extrapolation is shown by a dotted line):

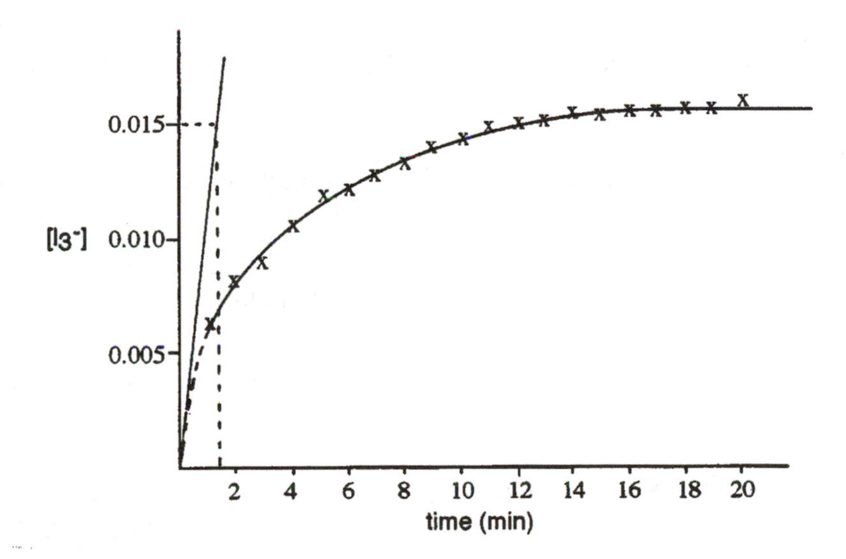

We can now begin to analyze the data properly.
For all of your data, we want to know the rate law, which will be of the form shown in equation 3:

$$\text{Rate} = k[S_2O_8^{2-}]^a[I^-]^b \tag{3}$$

We know the initial concentrations of all reactants (simply use the starting concentrations and volume and diluted volumes as we did above to get the initial $[S_2O_8^{2-}]$). If we can determine the rate, this will put us in a situation where we can determine k and a. (We will not bother to determine b as the experiment is over long as it is!)

One can define rate as being the Δ (concentration) ÷ Δ (time). We can determine the $\Delta [I_3^-] \div \Delta$ (time) using our graph. To get the initial rate, draw a tangent to the curve at time zero. The slope of this line is equal to the rate (see my graph for an example of this).

Thus, in my experiment, rate = $\Delta[I_3^-] \div \Delta(\text{time}) = (0.015\ M) \div (1.4\ \text{min}) = 0.0107\ M/\text{min}$

You should do this for all four of your runs, and complete the table in line *9* of the data sheet. I have done this for my data (for space reasons, I have not included all the data for runs 2 through 4):

Run	initial $[S_2O_8^{2-}]$	initial $[I^-]$	rate from graph (M/min)
1	0.0175	0.15	0.0107
2	0.00875	0.15	0.0055
3	0.0175	0.15	0.0221
4	0.0175	0.15	0.0462

Using the data in the table, you can calculate a variety of parameters:

i) Using the data from runs 1 and 2, calculate the order of reaction with respect to $S_2O_8^{2-}$ (the value of exponent a in general rate law equation, equation 3):

My data (prove this to yourself) show that a = 1. Put your value in line *10* of the data sheet.

ii) Using the data from runs 1 and 2, and assuming that the reaction is first order in iodide ions (that is, b = 1 in equation 3), calculate the rate constant (k):

My data show that k = 4.08 for run 1 and 4.38 for run 2. Average = 4.23 (record your results in lines *11–13*).

iii) Using the data from run 3 and the order of reaction from (i) and (ii), calculate the rate constant at 37 °C.

My data show that k = 8.42 at 37 °C (record your value in line *14*).

iv) Using the data from runs 1 and 3 and the Arrhenius equation, calculate the activation energy (E_a) for the reaction:

According to Arrhenius, $k = A \times \exp(-E_a/RT)$.

When you have data at two different temperatures, k and T change, all the other terms stay the same. So:

$$k_1 = A \times \exp(-E_a/RT_1) \quad \text{and} \quad k_2 = A \times \exp(-E_a/RT_2)$$

These can be combined to give: $\ln(k_2/k_1) = -E_a/R[1/T_2 - 1/T_1]$ (T in K, E_a in J!)

Using the equation, E_a from my data is equal to 80,800 J (record your value in line *15*).

Determination of an Equilibrium Constant

INTRODUCTION

Contrary to what you learned in the first semester, chemical reactions are rarely as simple as "reactants becoming products". Rather, reactions can generally go "both ways". For example, consider the reaction below:

$$A(g) + 2\ B(g) \rightarrow C(g) + D(g)$$

In "first semester language", if we start with some A and some B, these will react together in a ratio of 1 A to 2 B's to produce C and D. However, as more C and D are made, they may react together to produce A and B again. In this situation, we replace the $\rightarrow$ by $\Leftrightarrow$ indicate the "bi-directionality" of the reaction:

$$A(g) + 2\ B(g) \Leftrightarrow C(g) + D(g)$$

From your knowledge of reaction kinetics ("rates") you know that the rate of a reaction depends on the concentrations of the reactants; thus, the rate of the "forward" reaction (*i. e.*, $A + 2B \rightarrow C + D$) depends on the concentrations of A and B, and the rate of the "reverse" reaction (*i. e.*, $C + D \rightarrow A + 2B$) depends on the concentrations of C and D. At the start of the reaction, therefore, when there is only A and B present, the forward reaction will be fast, and the reverse non-existent. As more A and B become C and D, however, the rate of the forward reaction will decrease, and that of the reverse will increase, until, at some point, the two rates become equal. At this point, the reaction is said to be in equilibrium.

The point at which equilibrium is attained is determined by the concentrations. If we use the rate arguments above, then equilibrium is reached when:

$$\text{rate(forward reaction)} = \text{rate(reverse reaction)} \qquad (1)$$

If we consider both the forward and reverse reactions to be elementary steps for convenience, then:

$$\text{rate(forward)} = \text{k(forward)}[A][B]^2 \qquad (2)$$

$$\text{rate(reverse)} = \text{k(reverse)}[C][D] \qquad (3)$$

Combining equations (1) – (3), then, at equilibrium:

$$\text{k(forward)}[A][B]^2 = \text{k(reverse)}[C][D]$$

$$\frac{\text{k(forward)}}{\text{k(reverse)}} = \frac{[C][D]}{[A][B]^2}$$

The ratio of these rate constants is another constant at a particular temperature and is called the equilibrium constant (K_c). Thus, for any given reaction, when the relative concentrations of the chemicals equal the equilibrium constant at the particular temperature, the reaction is in equilibrium.

A knowledge of the equilibrium constant for a reaction is essential if one is to optimize the reaction conditions. In today's experiment, you will determine the equilibrium constant for the reaction:

$$Fe^{3+}(aq) + SCN^{-}(aq) \Leftrightarrow Fe(NCS)^{2+}(aq)$$

for which the equilibrium constant is,

$$K_c = \frac{[Fe(NCS)^{2+}]}{[Fe^{3+}][SCN^{-}]}$$

In order to determine the equilibrium constant, therefore, you need to measure the concentrations of each substance present at equilibrium. This is more easily said than done, however. There are generally two ways to measure the concentration of a substance—either remove the substance and see how much you have, or measure a concentration-dependent property of the substance. We cannot use the first method in this case as, once we remove a small amount of the substance, the reaction is no longer at equilibrium! We are, therefore, left with only the second option. Fortunately, the "product" of this reaction ($Fe(NCS)^{2+}$) has a deep-red color, so we may use spectrophotometric techniques to determine its concentration.

Thus, in a similar fashion to last week's lab, you should measure the transmittance of a number of reaction mixtures, convert these values to absorbance, and use the absorbance to calculate the concentration of $Fe(NCS)^{2+}$. This value may then be used to calculate K_c.

Easy, right? Not really, as, if you think about it a little, it is rather difficult to obtain the value for εl. Last week, you took a solution of known concentration, determined the absorbance and thus derived εl from Beer's Law. However, the fact that the reaction this week is reversible creates problems. For example, if you make a solution that starts as 0.1 M in $Fe(NCS)^{2+}$, it will immediately begin to break down into Fe^{3+} and SCN^{-} ions in order to achieve equilibrium. Thus, when you measure the transmittance of the solution, it will no longer correspond to a concentration of 0.1 M! In order to circumvent this problem you will do two things. You will do the reaction using a large excess of Fe^{3+} as compared to SCN^{-} ions. Under these conditions, it is a good assumption that nearly all of the SCN^{-} ions will react. Thus, $[Fe(NCS)^{2+}]$ at equilibrium will be the same as the starting $[SCN^{-}]$. (You can prove this to yourself if you pretend that K_c for the reaction is 400. If you start with 20 M Fe^{3+} and 0.2 M SCN^{-}, then, at equilibrium, $[Fe^{3+}]$ = 19.900101 M, $[SCN^{-}]$ = 0.000101 M and $[Fe(NCS)^{2+}]$ = 0.199899 M, which is about 0.2!). Secondly, rather than relying on only one measurement to derive εl, you will do the reaction several times with different concentrations of SCN^{-}. If you then plot a graph of concentration against absorbance, the slope of this graph will be εl. More importantly, you can use this graph to convert directly from absorbance values to concentrations.

Procedure

A. Preparation of Calibration Curve

1. Turn on the spectrophotometer.
2. Using a different, clean pipet for each solution, add the following amounts of solutions into seven clean beakers (note that the total volume in each beaker will be 10.0 mL). Stir thoroughly.

Beaker	0.002 M NaSCN in 0.1M HNO_3 (mL)	0.200 M $Fe(NO_3)_3$ in 0.1 M HNO_3 (mL)	0.1 M HNO_3 (mL)
1	0.0	2.5	7.5
2	0.1	2.5	7.4
3	0.2	2.5	7.3
4	0.4	2.5	7.1
5	0.6	2.5	6.9
6	0.8	2.5	6.7
7	1.0	2.5	6.5

3. Set the spectrophotometer to 447 nm (the wavelength of maximum absorption). Without a test tube in the sample holder, adjust the zero transmittance.
4. Pour some of the solution from beaker 1 into your cuvette, and use this to set the 100% transmittance. You should note that solution 1 has no SCN^-, thus is pure, yellow Fe^{3+}. By setting the 100% transmittance to this, you are effectively subtracting the effect of the Fe^{3+} from your measurements.
5. Clean the cuvette and then pour some of solution 2 into it and record its transmittance (*1a*).
6. Clean the cuvette thoroughly, and rinse with the next solution.
7. Repeat steps 5 and 6 with each of the remaining solutions.
8. Do the appropriate calculations and draw the calibration curve.

B. Collection of Equilibrium Data

9. In six clean, dry beakers, prepare the following reaction mixtures. Note that the concentration of the $Fe(NO_3)_3$ solution used in this part is much smaller than that used in Part A. Stir for about a minute.

Beaker	0.002 M NaSCN in 0.1M HNO_3 (mL)	0.002 M $Fe(NO_3)_3$ in 0.1 M HNO_3 (mL)	0.1 M HNO_3 (mL)
1	0	5	5
2	1.0	5.0	4.0
3	2.0	5.0	3.0
4	3.0	5.0	2.0
5	4.0	5.0	1.0
6	5.0	5.0	0.0

10. Recalibrate the 100% transmittance as in step 4 using the new beaker 1, which only contains the $Fe(NO_3)_3$ solution.
11. Rinsing the cuvette before each use with the appropriate solution, measure the transmittance of each reaction mixture (*2a*).
12. Carry out the appropriate calculations and determine your average K_c (*3*).
13. Obtain the class data (*4*) and calculate the class mean (*5*) and standard deviation (*6*).

Sample Calculations

A. *Preparation of Calibration Curve*

When I did the experiment, I obtained the following data:

1.	Beaker	1	2	3	4	5	6	7
a.	% Trans.	100	85	72	53	38	28	20

To calculate the absorbance (A), use the equation from last week's lab:

$$A = \log(100 \div \%T)$$

1.	Beaker	1	2	3	4	5	6	7
a.	% Trans.	100	85	72	53	38	28	20
b.	Absorb.	0	0.071	0.143	0.276	0.420	0.553	0.699

Finally, we need the equilibrium $[Fe(NCS)^{2+}]$. From our approximation,

$$\text{equilibrium } [Fe(NCS)^{2+}] = \text{initial } [SCN^-]$$

To get the initial $[SCN^-]$, simply use our favorite dilution equation,

$$\text{new concentration} = (\text{old concentration} \times \text{old volume}) \div \text{new volume}$$

Thus, for solution 2,

$$\text{new } [SCN^-] = (0.002 \times 0.1 \text{ mL}) \div 10 \text{ mL} = 2 \times 10^{-5} \text{ M:}$$

1.	Beaker	1	2	3	4	5	6	7
a.	% Trans.	100	85	72	53	38	28	20
b.	Absorb.	0	0.071	0.143	0.276	0.420	0.553	0.699
c.	$[Fe(NCS)^{2+}]$	0	2×10^{-5}	4×10^{-5}	8×10^{-5}	1.2×10^{-4}	1.6×10^{-4}	2×10^{-4}

Plotting the calibration curve:

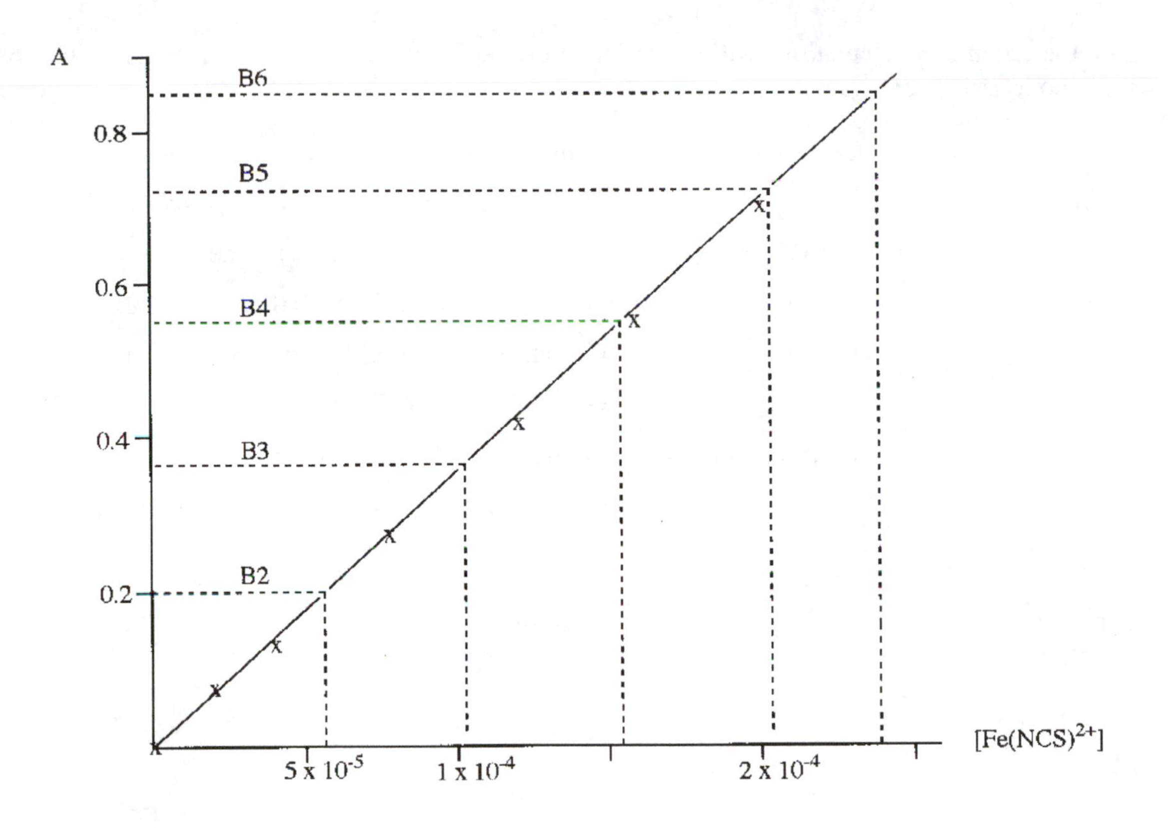

B. Collection of Equilibrium Data

When I did the experiment, I obtained the following data:

2.	Beaker	2	3	4	5	6
a.	% Trans.	64	43	28	19	14

You can calculate the absorbance as usual. Then, using the calibration curve, you can determine the equilibrium concentration of $Fe(NCS)^{2+}$ (shown on the calibration curve with the dotted lines):

2.	Beaker	2	3	4	5	6
a.	% Trans.	64	43	28	19	14
b.	Absorb.	0.194	0.367	0.553	0.721	0.854
c.	$[Fe(NCS)^{2+}]$	5.5×10^{-5}	1.05×10^{-4}	1.58×10^{-4}	2.06×10^{-4}	2.44×10^{-4}

Remember that all of the concentrations at equilibrium are required to determine K_c; thus, the equilibrium values of $[Fe^{3+}]$ and $[SCN^-]$ are still needed. This may be done using stoichiometry, since one mole of $Fe(NCS)^{2+}$ is made from one mole of Fe^{3+} and one mole of SCN^-. Thus, the amount of Fe^{3+} and SCN^- used to attain equilibrium is the same as the amount of $Fe(NCS)^{2+}$ made:

$$\text{equilibrium } [Fe^{3+}] = \text{initial } [Fe^{3+}] - \text{equilibrium } [Fe(NCS)^{2+}]$$

$$\text{equilibrium } [SCN^-] = \text{initial } [SCN^-] - \text{equilibrium } [Fe(NCS)^{2+}]$$

This, and the remaining calculations will be carried out for solution 2, and then the final chart will be shown for your assistance and edification!

Part B Calculation Equations

Initial $[Fe^{3+}]$ $= (\text{old } [Fe^{3+}] \times \text{old volume}) \div (\text{new volume})$

$= (0.002\ M \times 5.0\ mL) \div 10.0\ mL =$ **0.001 M**

Initial $[SCN^-]$ $= (\text{old } [SCN^-] \times \text{old volume}) \div (\text{new volume})$

$= (0.002\ M \times 1.0\ mL) \div 10.0\ mL =$ **0.0002 M**

Equilibrium $[Fe^{3+}]$ $= \text{initial } [Fe^{3+}] - \text{equilibrium } [Fe(NCS)^{2+}]$

$= 0.001\ M - 5.5 \times 10^{-5}\ M$ $=$ **9.45×10^{-4} M**

Equilibrium $[SCN^-]$ $= \text{initial } [SCN^-] - \text{equilibrium } [Fe(NCS)^{2+}]$

$= 0.0002\ M - 5.5 \times 10^{-5}\ M$ $=$ **1.45×10^{-4} M**

K_c

$$= \frac{[Fe(NCS)^{2+}]}{[Fe^{3+}][SCN^-]}$$

$$= \frac{5.5 \times 10^{-5}}{(9.45 \times 10^{-4})(1.45 \times 10^{-4})}$$

= 401

2.	Beaker	2	3	4	5	6
a.	% Trans.	64	43	28	19	14
b.	Absorb.	0.194	0.367	0.553	0.721	0.854
c.	$[Fe(NCS)^{2+}]$	5.5×10^{-5}	1.05×10^{-4}	1.58×10^{-4}	2.06×10^{-4}	2.44×10^{-4}
d.	Init. $[Fe^{3+}]$	0.001	0.001	0.001	0.001	0.001
e.	Init. $[SCN^-]$	0.0002	0.0004	0.0006	0.0008	0.001
f.	Eqm. $[Fe^{3+}]$*	9.45×10^{-4}	8.95×10^{-4}	8.42×10^{-4}	7.94×10^{-4}	7.56×10^{-4}
g.	Eqm. $[SCN^-]$	1.45×10^{-4}	2.95×10^{-4}	4.42×10^{-4}	5.94×10^{-4}	7.56×10^{-4}
h.	K_c	401	398	425	437	427

*Eqm. stands for equilibrium

Introduction to Acids and Bases

Introduction

The words "acid" and "base" are ones that you have undoubtedly heard throughout your education. In this experiment, you will observe some of the traditional properties of acids and bases, as well as their importance in our everyday lives.

The lab is divided into three parts:

i) In the first section of the lab, you will compare the properties of an acid (0.1 M HCl) and a base (0.1 M NaOH) with those of water.

ii) In the second section of the lab, you will "mimic" one form of acid rain production and examine its properties. Acid rain is a major problem facing society today. The phenomenon occurs when gases produced by various societal or industrial activities react with water in the atmosphere to produce acids. For example, when sulfur burns in air, it produces $SO_2(g)$ (equation 1). This gas can react with water to produce sulfurous acid (H_2SO_3) (equation 2). This is the acid which you will be investigating. In the atmosphere, however, sulfurous acid is slowly oxidized to sulfuric acid! (equation 3).

$$S + O_2(g) + \text{heat} \rightarrow SO_2(g) \qquad (1)$$

$$SO_2(g) + H_2O(l) \rightarrow H_2SO_3(aq) \qquad (2)$$

$$2\ H_2SO_3(aq) + O_2(g) \rightarrow 2\ H_2SO_4\ (aq) \qquad (3)$$

iii) In the third part of the experiment, you will determine the acidity or basicity of several common substances. There are two purposes to this exercise:

a) to realize how prevalent acids and bases are in everyday life,

b) to identify which "ingredient" in each substance gives rise to the acid/base properties.

Procedure

A. Comparison Between Acids, Bases, and Water

In the first section of the lab, you will compare the properties of an acid (0.1 M HCl) and a base (0.1 M NaOH) with those of water. You will perform the same "test" on each of the three substances before moving to the next test.

1. (*Optional*). Wash a beaker thoroughly and pour in some of the sample. Take a SMALL mouthful and observe and record the taste (*1*). Spit the mouthful out (gracefully). DO NOT SWALLOW!

2. Dip your fingers into a beaker of each sample and record the feel (*2*). Rinse your fingers well.
3. Add about 5 mL of a sample to a test tube. Add a 1 cm strip of magnesium ribbon, observe and record (*3*).
4. Add about 5 mL of a sample to a test tube. Add a small piece of calcium carbonate and record your observations (*4*).
5. Add about 5 mL of a sample to a test tube. Add a small piece of ammonium chloride and smell carefully, using the correct procedure. Record your observations (*5*).
6. Add about 5 mL of a sample to a beaker. Hold a piece of "red litmus paper" in the sample and record your observations (*6*).
7. Add about 5 mL of a sample to a beaker. Hold a piece of "blue litmus paper" in the sample and record your observations (*7*).
8. Pipet 5.0 mL of HCl and add it to a beaker. To this beaker add 5.0 mL NaOH. Stir the mixture and repeat steps 1 through 7. Explain your results (*8*).

B. Acid Rain

DO THIS IN THE HOOD

9. Set up a triangle supported on a ring.
10. Place a crucible in the triangle and begin to heat.
11. Weigh about 2 g of sulfur and place into the hot crucible. Allow the sulfur to ignite.
12. Using tongs, hold a large test tube inverted directly above the crucible to collect the gas.

THE GAS PRODUCED IS HIGHLY TOXIC—DO NOT INHALE!

13. When all the sulfur has finished burning, stopper the test tube while it is still inverted. Allow the test tube to cool.
14. Turn off the burner and place a lid on the crucible. Allow to cool.
15. When the test tube is cool, add 5 mL of water, re-stopper and shake to dissolve the gas.
16. At your bench area, divide the contents of the tube into two "medium" test tubes.
17. Using blue litmus paper, test the acidity of the contents of one test tube (*9*).
18. Add a small piece of Mg ribbon to this tube and record your observations (*10*).
19. Add a small piece of calcium carbonate to a second tube and record your observations (*11*).

C. *Acids and Bases at Home*

In order to determine the pH of the substance, you will use a pH meter. While we could just use litmus paper as before, many of these substances are colored, which can confuse the issue.

A pH meter consists of two parts—the meter and the electrode. The electrode is made of glass—SO BE CAREFUL NOT TO BREAK IT. It can also dry out, so ALWAYS KEEP IT IN LIQUID (unless making a measurement, a beaker of water is sufficient). Each pH meter is accompanied by two beakers of colored buffer solutions—a yellow one (pH 7) and a red one (pH 4). These will be used to calibrate the pH meter. You also should have a beaker of water or a wash bottle which you should use to rinse and store the electrode.

20. Rinse the electrode, either by swirling in a beaker of water, or, preferably, by squirting water from a wash bottle GENTLY up and down the probe.
21. Turn the meter on. If necessary (it shouldn't be) adjust it to read pH (if unsure, ask your TA).
22. Press the CAL button. After a little while, the meter will display the last calibration used. This should be "7-4". If it is, press the YES key. If it is not, press either ▲ or ▼until the screen reads "7-4". Press the YES key.
23. Place electrode in pH 7 buffer. When READY light comes on, press the YES key.
24. Remove electrode from pH 7 buffer. Rinse well and place in the pH 4 buffer.
25. When READY light comes on, press the YES key.
26. Rinse the electrode. The pH meter is now ready to use.
27. There are several commercial substances laid out for you in beakers (listed on data sheet). Test the pH of each using the pH meter by simply placing the top of the electrode in the sample. When the reading has stabilized, record the value (*12*). Rinse the electrode well before measuring each sample. You might want to check the calibration of the meter every 3 or 4 samples (steps 22 through 26).
28. Identify each substance as acid or base, then, during the week, find out which "ingredient" in each is responsible for the acidity or basicity of the substance. Record this in (*12*) of the data sheet.

Acid-Base Titration Curves Using a pH Meter

INTRODUCTION

Titration is a technique used to add incrementally precise volumes of a liquid (usually using a buret). In this experiment, you will add small amounts of a base to different acids, and measure the effect of each addition on the pH of the acid.

In any acid-base titration, if the base is added to the acid, the pH will increase slightly until the equivalence point is reached. At this point, the pH will increase dramatically as all of the acid has been neutralized. Further addition of base after this point simply increases the pH gradually. This form of behavior is shown below in a graph of pH vs. volume of base added (Fig. 17–1):

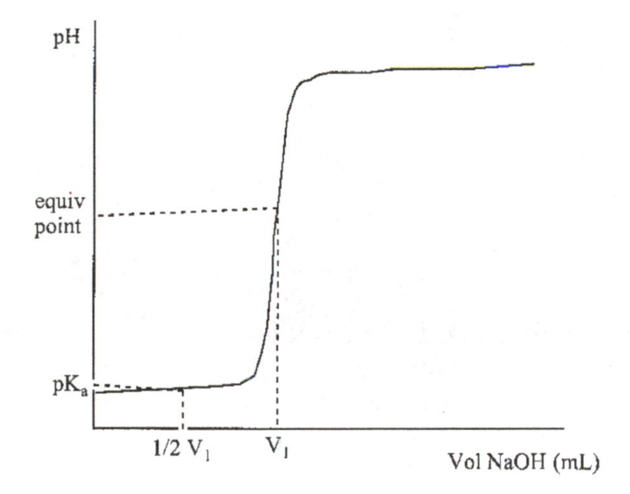

Figure 17–1

The actual pH of the equivalence point depends upon the nature of the acid and base used. When a strong acid (such as HCl) and a strong base (such as NaOH) are used, neutralization yields water and the spectator ions, Na^+ and Cl^-. At the equivalence point, therefore, the pH is 7. When a weak acid (such as acetic acid) is titrated by a strong base, however, neutralization yields water, the spectator ion, Na^+, and the conjugate base of the weak acid (acetate). At the equivalence point, therefore, there is a weak base present in the solution, so the pH will be greater than 7.

We can generalize this situation:

for any acid (HA), we may write the following equilibrium:

$$HA(aq) \Leftrightarrow H^+(aq) + A^-(aq) \qquad (1)$$

and the corresponding acid-dissociation constant expression is:

$$K_a = \frac{[H^+][A^-]}{[HA]} \qquad (2)$$

Equation (2) may be rearranged to give:

$$[H^+] = \frac{K_a[HA]}{[A^-]} \qquad (3)$$

Taking negative logs of both sides and rearranging slightly,

$$-\log [H+] = -\log K_a - \log \frac{[HA]}{[A^-]} \qquad (4)$$

or, using the fact that $pH = -\log[H^+]$ and $pK_a = -\log K_a$,

$$pH = pK_a - \log \frac{[HA]}{[A^-]} \qquad (5)$$

In the titration of a weak acid with a strong base, there is obviously a point at which half of the acid has been neutralized to the conjugate base, and half remains protonated, i. e., $[HA] = [A^-]$, or $[HA]/[A^-] = 1$. At this point, therefore, from equation (5),

$$pH = pK_a - \log 1$$

or,

$$pH = pK_a \qquad (6)$$

A graph of pH vs. volume of base added tells us the amount of base required to reach the equivalence point (*i. e.*, the amount of base needed to neutralize all of the acid). Therefore, the volume at the half-equivalence point (the point at which equation 6 is valid) is simply half of the volume at the equivalence point. Thus, by plotting pH vs. volume of base, determining the amount of base required to reach equivalence, dividing that amount of base by two and reading the corresponding pH from the curve provides a convenient method for determining the pK_a (and thus the K_a) of the acid.

It is often easier to plot a second form of graph, as the exact equivalence point is hard to determine graphically. Plotting the change in pH/change in volume vs. the average volume provides a sharp peak at the equivalence point, as shown in Figure 17–2.

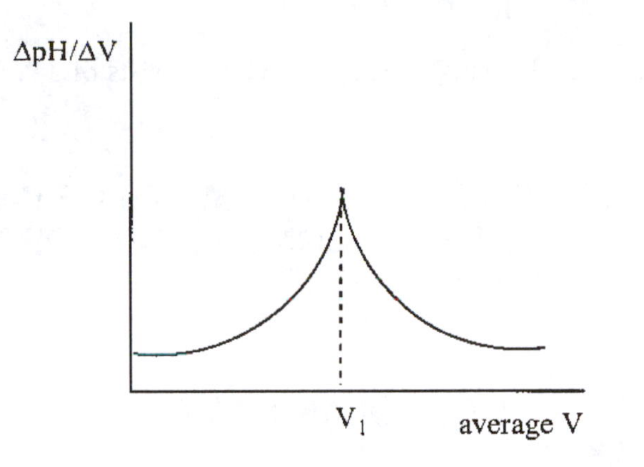

Figure 17–2

We can also apply this approach to the study of a polyprotic acid, H_2A. Such acids have more than one acidic hydrogen, each of which has a different associated K_a. A pH titration curve of such an acid, therefore, has more than one equivalence point (Figure 17–3).

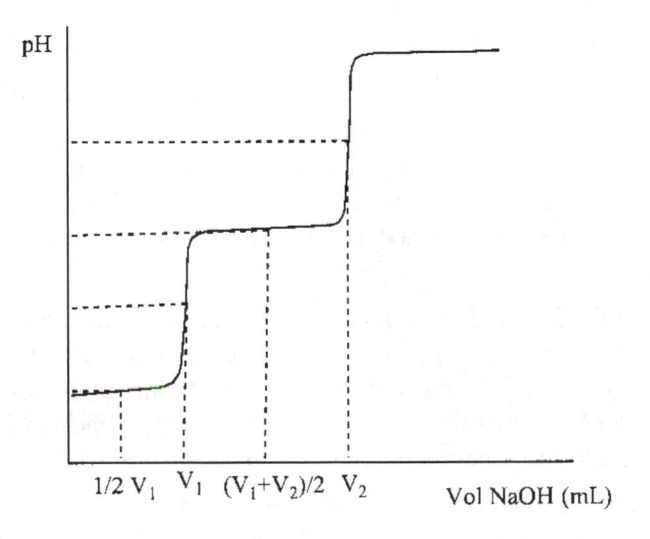

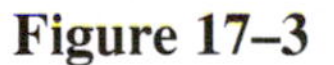

Figure 17–3

In this experiment, you will carry out titrations of three different acids, a strong acid, a weak acid, and a polyprotic acid. Each time, you should measure the pH changes as you add sodium hydroxide. For each titration, you should plot two graphs—one of pH vs. volume (Figure 17–1) and the one of (change in pH/change in volume) vs. average volume (Figure 17–2). From the second type of graph, you will determine the pH of the equivalence point(s) for each acid. Finally, for the weak and polyprotic acids, you will determine the values of K_a from the pH at the half-equivalence points.

PROCEDURE

In this lab we will measure pH using a pH meter. This consists of a "box" and an electrode. Two things should be noted immediately.

THE ELECTRODE SHOULD ALWAYS BE IMMERSED IN A LIQUID — EITHER YOUR REACTION MIXTURE OR THE BUFFER SOLUTION PROVIDED.

THE ELECTRODE IS GLASS SO CARE SHOULD BE TAKEN NOT TO BREAK IT.

A. Titration Curve of a Strong Acid

1. When you first go to your work station, you will see a pH meter, an electrode, and two beakers, one with a yellow solution (a buffer of pH 7) and a red solution (a buffer of pH 4). Use these to calibrate the pH meter as described in Experiment 16.
2. Take a beaker to the NaOH carboy and obtain about 100 mL standarized NaOH solution (record the exact concentration in line *1* of the data sheet). Fill the clean, rinsed buret from the beaker.
3. Take another beaker, and pipet 25.0 mL of 0.1 M HCl into it. Record the exact volume on line *2* of the data sheet.
4. Now place the acid beaker under the buret and put the pH electrode into the solution (obviously, the electrode should be still connected to the meter!) and stir the solution (being careful not to hit the electrode). Once the pH is stable (*i. e.*, not changing within a space of 4–5 seconds) record this in line *3a*.
5. Add 2 mL of NaOH, stir, and again record the pH after stabilization (*3b*). Continue to add 2 mL increments until the equivalence point is near. You can tell when this is happening as the pH starts to increase in a more dramatic fashion. At this point, begin adding NaOH in increments of 0.2 mL. Once the pH change begins to flatten off again, go back to 2 mL increments for four more readings. Each time, record the pH and volume.
6. Plot a graph of pH against volume of base. Calculate the values of "ΔpH", "ΔV", "ΔpH/ V" and "avg. vol." in the data sheet. Plot a graph of "ΔpH/V" against "avg. vol."
7. Determine the volume of and the pH at the equivalence point (*4, 5*).

B. Titration Curve of a Weak Acid

8. Check the calibration of the pH meter as in step 1.
9. Take a clean beaker, and pipet 25.0 mL of 0.1 M of the unknown weak acid into it. Record the exact volume on line *6* of the data sheet.
10. Repeat steps 4 though 7 above using lines *7* through *9* on the data sheet.
11. Determine the volume and pH at the half-equivalence point (*10* and *11*) and thus the K_a of your acid (*12*).

C. *Titration Curve of a Polyprotic Acid*

12. Check the calibration of the pH meter as in step 1.
13. Take a clean beaker, and pipet 25.0 mL of 0.1 M of the unknown polyprotic acid into it. Record the exact volume on line *13* of the data sheet.
14. Repeat steps 4 though 6 above using line *14* on the data sheet.
15. Determine the volume and pH for each equivalence point (*15* and *16*).
16. For each of these, determine the volume and pH at the half-equivalence point (*17* and *18*) and thus the values of the different K_a's (*19*).

SAMPLE CALCULATIONS

I will use the weak acid titration as the example:

Plot the pH against volume to produce a plot such as Figure 17–1.

Fill out the rest of the columns in line *7* of the data sheet:

· Column 3 (ΔpH) is simply the measured pH minus the pH from the previous line

· Column 4 (ΔV) is the difference in volume between the two steps

· Column 5 (ΔpH /ΔV) is obvious

Column 6 (avg. Vol.) is the average of (the measured volume plus the previous one) *e. g.*

C1 Volume	C2 pH	C3 ΔpH	C4 ΔV	C5 ΔpH/ΔV	C6 avg. vol.
0	2.78	0	-	0	0
2	2.89	0.11	2	0.055	1
4	2.98	0.09	2	0.045	3
6	3.12	0.14	2	0.07	5
8	3.24	0.12	2	0.06	7
10	3.40	0.16	2	0.08	9

Plot column 5 vs. column 6 to get a curve such as Figure 17–2.

Determine the volume at the equivalence point from this second curve (line *8*).

Determine the volume of the half-equivalence point (line *10*).

Read the pH at this volume from the first graph (line *11*).

Calculate the value of K_a (line *12*).

See chem II
notes prof
chem

Slightly Soluble Salts

INTRODUCTION

The idea of "solubility" is a highly important one in our lives. You are probably well-acquainted with very soluble chemicals (salt and sugar in water, for example). While less familiar, however, the concept of slightly soluble compounds is of equal importance (for example, many of the problems of water pollution are due to the presence of slightly soluble chemicals).

The difference between a "soluble" and "insoluble" (or "slightly soluble") compound is, obviously, a question of how much of a compound can be dissolved in water. For example, at 25 °C, one can dissolve nearly 36 g of NaCl in 100 mL of water but only 8.9×10^{-5} g of AgCl. NaCl is "soluble", AgCl is "insoluble". In today's experiment, you will work with "insoluble" salts in two different ways. First, you will quantify the solubility of such a salt, using a titration. Secondly, you will examine briefly one application of salt insolubility—qualitative inorganic analysis.

Determination of K_{sp}

In order to quantify the "insolubility" of salts such as AgCl, we cannot directly measure how much can be dissolved (as it is such a small amount)—we must take a roundabout route. The easiest method, for us, is to use the idea of dynamic equilibrium. For any salt, if one adds more salt to water than can be dissolved, you obtain a heterogeneous system, in which a saturated solution of the salt (a solution in which no more salt can be dissolved) is in contact with the undissolved, excess salt. In such a case, the ions in solution are continually solidifying at the same rate as previously "solid" ions are dissolving, resulting in a state of equilibrium. In this state, the concentrations of the ions in solution are constant, and we can use our normal equilibrium concepts to examine the solution.

For example, let us consider taking a saturated solution of the "insoluble" salt $PbCl_2$ in contact with the solid. The following equilibrium will be in force:

$$PbCl_2(s) \Leftrightarrow Pb^{2+}(aq) + 2\ Cl^{-}(aq)$$

As this is an equilibrium, we can set up the mass action expression, which, in this special case refers to a solubility product equilibrium constant, the K_{sp}:

$$K_{sp} = [Pb^{2+}][Cl^{-}]^2$$

Note that the solid $PbCl_2$ does not appear in this expression.

As K_{sp} is a constant at a given temperature, we may define the "solubility" of an "insoluble salt" using the value of K_{sp}, and we may derive the value of K_{sp} by measuring the concentrations of the ions present in a saturated solution. (You should note, that, as all ions come from the same source, all you have to do is determine the concentration of one of the ions).

In the first part of the experiment, you are going to calculate the K_{sp} value of an "insoluble" hydroxide—$Ca(OH)_2$. In solution, this dissociates into calcium ions and hydroxide ions, thus providing us with an easy way of determining the concentrations—an acid titration.

Qualitative Analysis

Chemical analysis may be divided into two broad types—quantitative and qualitative. You have generally used quantitative analysis—you have determined "how much?" of a substance. In the second part of this lab, you will simply determine "what?" In particular, you will apply solubility chemistry to determine qualitatively which ions are present in a particular solution.

The traditional approach to qualitative analysis of ions divides different cations or anions into different groups, in which members of the same group all form the same type of insoluble salt. For example, Group I cations all form insoluble chlorides. In today's experiment, you will be interested in the following, specific groups:

Cations		Anions	
Group 1	Group II	Group I	Group II
Silver (Ag^+)	Magnesium (Mg^{2+})	Chloride (Cl^-)	Carbonate (CO_3^{2-})
Lead (Pb^{2+})	Calcium (Ca^{2+})	Bromide (Br^-)	Sulfate (SO_4^{2-})
Mercury(I) (Hg_2^{2+})	Barium (Ba^{2+})	Iodide (I^-)	Phosphate (PO_4^{3-})

Thus, Group I cations have insoluble chlorides, and Group II cations form insoluble phosphates (among others). Group I anions are the halides, and have insoluble silver salts, while Group II anions form insoluble salts with calcium and barium. In order, therefore, to distinguish cation Group I, one adds HCl to a sample. If a Group I cation is present in the sample, a precipitate will form. No precipiate implies that a Group I cation is not present. Once the group has been identified, there are tests run within each group that help to determine the exact identity of the ion present. In today's experiment, you will be given an "unknown" solution, that either contains an unknown cation or an unknown anion (with the appropriate counter ions). You will be required to determine the identity of your unknown ion!

PROCEDURE

A. Determination of K_{sp}

1. Rinse the buret in the usual way with 0.05 M HCl solution. Record the exact concentration of the HCl on the data sheet (*1*). Fill the buret with the HCl solution.
2. Decant some (about 55 mL) of the saturated $Ca(OH)_2$ solution into your beaker. Be sure that you do not get any of the solid also.
3. Pipet about 25 mL $Ca(OH)_2$ solution from your beaker into an Erlenmeyer flask. Record the exact volume on the data sheet (*2*). Add a couple of drops of methyl orange indicator (methyl orange is yellow in basic solution and red in acidic).
4. Record the initial volume of the HCl in the buret (*3*) and titrate the $Ca(OH)_2$ until you see the color change. Record the final volume (*4*).
5. Repeat steps 3 and 4.

6. Do appropriate calculations to determine your K_{sp} (*5 –10*)
7. Using class data (*11*), calculate the class mean (*12*) and standard deviation (*13*).

B. Qualitative Analysis

8. See your TA for about 5 mL of your unknown solution. Record which unknown you were given, and whether it is a cation or anion unknown (*14*).
9. Use the method listed in the flow charts (Figure 18–1 for cations, 18–2 for anions) to determine your unknown. Record all observations in the appropriate place on the data sheet (*15*), and identify your unknown ion (*16*).

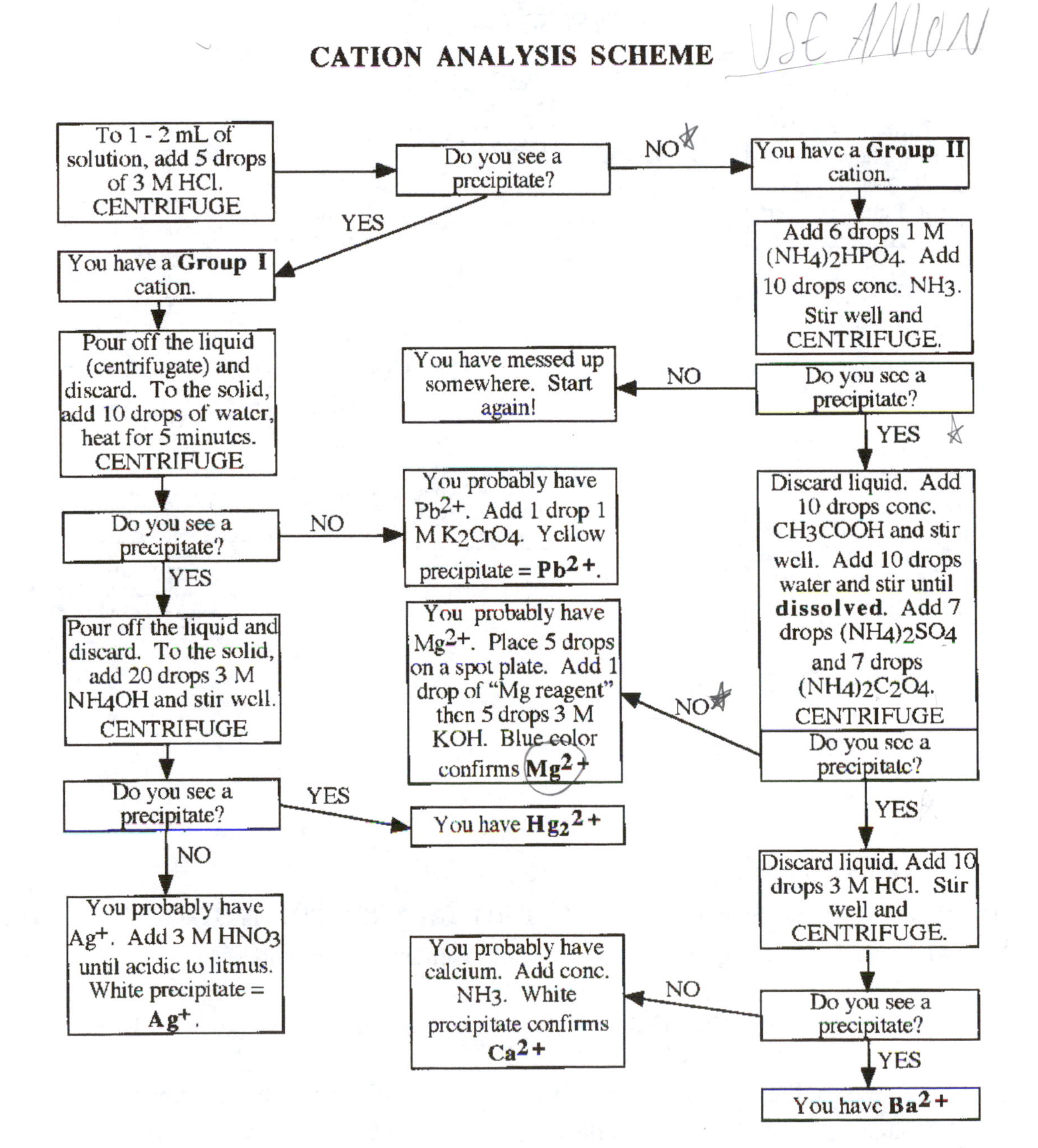

ANION ANALYSIS SCHEME

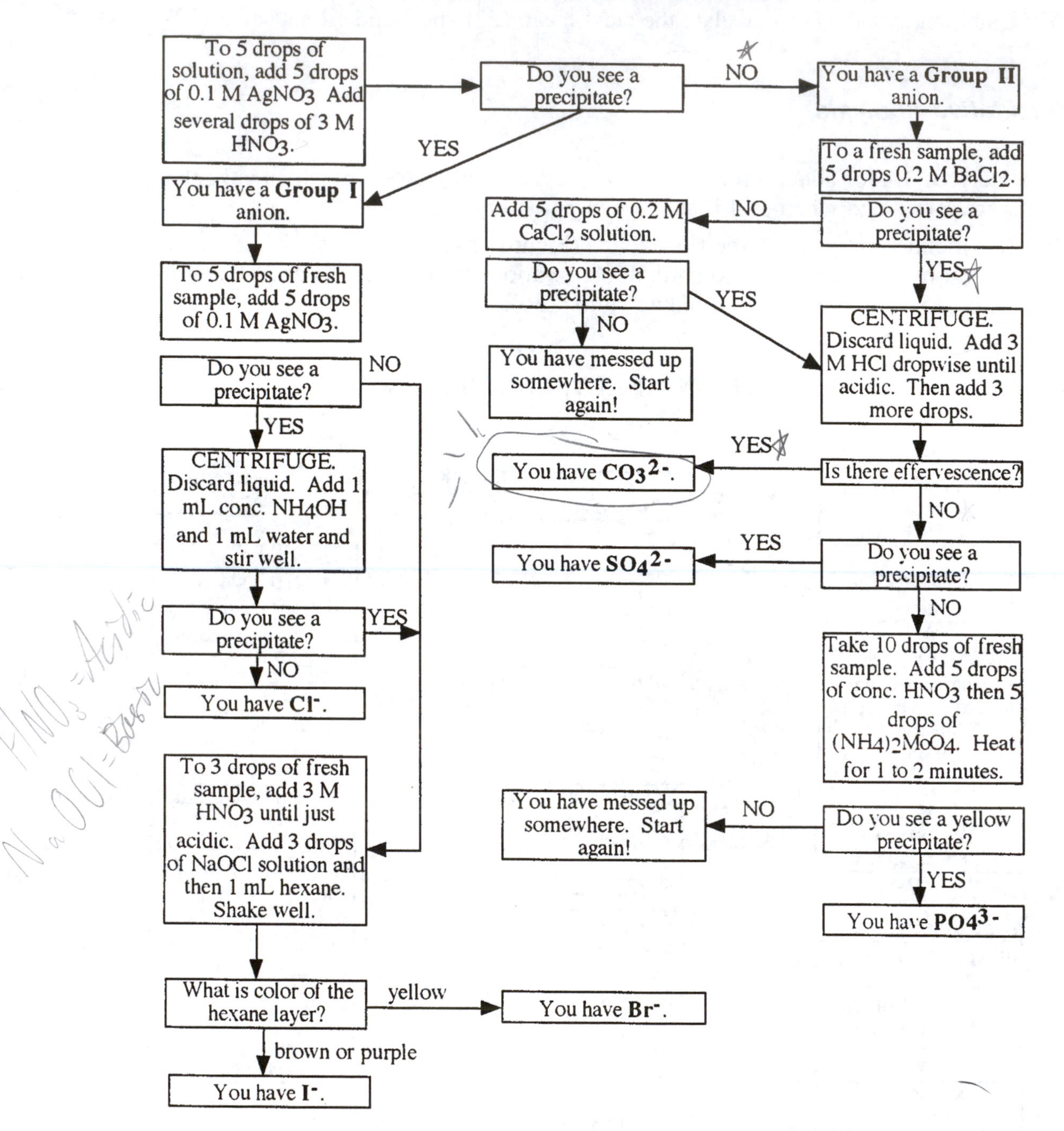

Some important points:

10. When using the centrifuges, BE SURE TO BALANCE THE TEST TUBES INSIDE.
11. Always use CLEAN, DRY TEST TUBES. Contamination will seriously mess up your results!
12. There are a lot of different chemicals in use. CHECK AND DOUBLE CHECK that you are using the correct one.
13. You are only looking for the presence of one ion. When you have found it, you can stop!
14. If you are unsure about the results of a test, there are known solutions of each possible ion in the laboratory. Use these to see what should happen for a particular test!

Sample Calculations

We are going to calculate the value of K_{sp} for $Ca(OH)_2$. The equilibrium, in effect, in our saturated solution in contact with the solid is:

$$Ca(OH)_2(s) \Leftrightarrow Ca^{2+}(aq) + 2\ OH^-(aq)$$

which produces the equation:

$$K_{sp} = [Ca^{2+}][OH^-]^2$$

Thus, we need to know the concentrations of both Ca^{2+} and OH^- in a saturated solution.
All of the ions, however, came from the same source - solid $Ca(OH)_2$. Thus, according to stoichiometry,

$$[Ca^{2+}] = 1/2 \cdot [OH^-]$$

We can, therefore rewrite the K_{sp} equation:

$$K_{sp} = 1/2 \cdot [OH^-][OH^-]^2$$

So, all we have to do is determine the concentration of OH^- in the solution and we are set.

The method we used to determine the concentration of OH^- was titration to the neutralization point with acid. Using the results of this titration, it should be routine for you by now to calculate all concentrations needed.

When I did the experiment, I obtained:

1. [HCl] = 0.0200 M
2. Volume of $Ca(OH)_2$ = 25.0 mL
3. Initial buret reading = 0.0 mL
4. Final buret reading = 20.0 mL
5. Volume of acid added = final buret reading – initial reading
   = 20.0 mL – 0.0 mL
   = 20.0 mL
6. Moles of acid added:
   molarity of acid = moles of acid / volume of acid (L)
   moles acid = molarity × volume (IN LITERS)
   = 0.0200 (moles/liter) × 0.0200 L
   = 4.00×10^{-4} moles
7. The moles of acid = moles of OH^- present, according to the neutralization equation:

$$H^+(aq) + OH^-(aq) \rightarrow H_2O(l)$$

Thus, our solution of saturated $Ca(OH)_2$ contained 4.00×10^{-4} moles of OH^-.

8. Having obtained moles of OH-, we need the concentration of OH-:

$$\text{molarity of } OH^- = \frac{\text{moles of } OH^-}{\text{volume of } OH^- \text{ (L)}}$$

$$= \frac{4.00 \times 10^{-4} \text{ moles } OH^-}{0.025 \text{ L}}$$

$$= 1.60 \times 10^{-2} \text{ moles/liter}$$

9. Now, the end is in sight. All we have to do is substitute the value of $[OH^-]$ into our equation for K_{sp}:

$$K_{sp} = [Ca^{2+}][OH^-]^2 = 1/2 \bullet [OH^-][OH^-]^2$$

$$= 1/2 \times (1.60 \times 10^{-2})(1.60 \times 10^{-2})^2$$

$$= \mathbf{2.05 \times 10^{-6}}$$

Electrochemistry I. Galvanic Cells

INTRODUCTION

By now, you should be familiar with the concepts of reduction and oxidation. In particular reactions, some species tend to lose electrons (are oxidized), others tend to gain electrons (are reduced). You should also be familiar with the idea that we can apply energetic considerations to *any* reaction and determine whether or not it is spontaneous. In this experiment, you will examine the spontaneity of some redox reactions, and express this quantitatively in terms, not of free energy or equilibrium, but of **reaction potential**. This value reflects the extent to which electrons can be transferred between the reacting species. Accordingly, it is measured in volts.

Rather than trying to consider an entire redox (or **electrochemical**) reaction, we will focus on the different halves of the reactions—the reduction **half-reaction** and the oxidation **half-reaction**. We can assign a potential to each half-reaction that reflects the tendency of the reacting species to gain or to lose electrons respectively. Thus, for example, the values given in the Table:

Half-reaction	Potential (E°) V
$Cl_2(g) + 2\ e^- \rightarrow 2\ Cl^-(aq)$	+1.36
$Br_2(l) + 2\ e^- \rightarrow 2\ Br^-(aq)$	+1.07
$Ag^+(aq) + e^- \rightarrow Ag(s)$	+0.80
$2\ H^+(aq) + 2\ e^- \rightarrow H_2(g)$	0.00
$Ni^{2+}(aq) + 2e^- \rightarrow Ni(s)$	-0.25
$Na^+(aq) + e^- \rightarrow Na(s)$	-2.71
$K^+(aq) + e^- \rightarrow K(s)$	-2.93

These values are listed as **reduction** potentials, and reflect the extent to which the reactant "wants" to gain electrons. The more positive the potential, the greater this "desire". Thus, chlorine atoms "want" to gain electrons to become chloride anions, while potassium cations really do not "want" to gain electrons to become potassium atoms. This is in agreement with the principles of reactivity that you have learned up to now.

We can also use these numbers to express oxidation potentials. As oxidation is the reverse of reduction, we would rewrite the table as follows. Note that, when we reverse the direction of the reaction, we also reverse the sign of the potential:

Half-reaction	Potential (E°) V
$2\ Cl^-(aq) \rightarrow Cl_2(g) + 2\ e^-$	-1.36
$2\ Br^-(aq) \rightarrow Br_2(l) + 2\ e^-$	-1.07
$Ag(s) \rightarrow Ag^+\ (aq) + e^-$	-0.80
$H_2(g) \rightarrow 2\ H^+(aq) + 2\ e^-$	0.00
$Ni(s) \rightarrow Ni^{2+}(aq) + 2e^-$	+0.25
$Na(s) \rightarrow Na^+(aq) + e^-$	+2.71
$K(s) \rightarrow K^+(aq) + e^-$	+2.93

In other words, sodium and potassium really "want" to lose an electron each to become cations, while chloride and bromide ions are very stable with respect to their elements.

One can use these half-reaction potentials to predict the spontaneity of a redox reaction, as the potential of the overall reaction is simply the sum of the half-reactions. In addition, one can also predict whether or not a given reaction would occur.

For example, you know that sodium cations and chloride anions will not react, but sodium metal and chlorine gas will react, rather vigorously. One can explain these observations rather easily using the half-reaction potentials:

For the reaction between Na and Cl, the half-reactions are:

		E°(V)
reduction:	$Na^+(aq) + e^- \rightarrow Na(s)$	-2.71 V
oxidation:	$2\ Cl^-(aq) \rightarrow Cl_2(g) + 2\ e^-$	-1.36 V

The overall potential for the reaction would be (-2.71) + (-1.36) or -4.07 V—a very negative number.

Conversely, the half-reactions for the reaction between sodium metal and chlorine gas are:

oxidation:	$Na(s) \rightarrow Na^+(aq) + e^-$	+2.71 V
reduction:	$Cl_2(g) + 2\ e^- \rightarrow 2\ Cl^-(aq)$	+1.36 V

The overall potential for this reaction would be (2.71) + (1.36) or +4.07 V—a very positive number.

The reaction between sodium and chlorine is rather easy to predict as sodium "wants" to lose an electron while chlorine "wants" to gain electrons. We can also use the potentials to predict how a reaction will proceed when we have species that want to do the same thing. For example, consider chlorine and bromine. Examination of the potentials shows that both elements want to be reduced, while the two derived anions, chloride and bromide, do not want to be oxidized. However, one can oxidize bromide if it is treated with chlorine:

reduction:	$Cl_2(g) + 2\ e^- \rightarrow 2\ Cl^-(aq)$	E = +1.36 V
oxidation:	$2\ Br^-(aq) \rightarrow Br_2(l) + 2\ e^-$	E = -1.07 V
		E(reaction) = +0.29 V

On the other hand, however, as the E° value associated with chlorine and chloride is higher than that associated with bromine and bromide, bromine will not oxidize chloride ions.

You examined this type of chemistry in Experiment 10, when you investigated the relative reactivity or "activity" of the halogens. You found that chlorine was more reactive than bromine, which, in turn, was more reactive than iodine. This activity series is matched by the E° values for the halogens.

In the first part of the experiment, you will apply exactly the same reasoning to a series of metals. Metal atoms generally tend to lose electrons, to form cations. Some metal atoms want to lose electrons more than others. We can determine the order of activity, therefore, by adding elemental metals to solutions that contain cations of other metals. If the added metal disappears, and the second one is produced from its cations, then the first metal is more reactive. It wants to lose electrons more than the second metal; therefore, it "forces" the cations of the second metal to accept its "spare" electrons. We can add an extra "element" into the list—hydrogen. You should know that hydrogen tends to form the cation, H^+. Therefore, if we add a metal to a solution of H^+ cations (an acid!) and it bubbles (that is, elemental hydrogen gas is formed), we can say that the metal is more active than hydrogen. On the other hand, if nothing happens, then the metal is less active than hydrogen.

With this activity series in hand, you will measure E° values for the some of the same metals (and some new ones) by creating **galvanic** cells. In such a cell, the components of two half-reactions are placed in two separate containers. The solids (electrodes) are connected by wires that are in turn connected to a voltmeter, while the two solutions are connected by a **salt bridge** (a piece of filter paper soaked with an ionic solution)—used to balance the charge of the system. Electrons are "lost" by a species in one container, travel *via* the electrode (the **anode**) through the wire and voltmeter to the electrode in the other container (the **cathode**) where they take part in the reduction. The reading measured on the voltmeter is reflective of the value of the overall E°.

Finally, you will use the observations from the first two parts to construct two types of "batteries" (devices that produce electricity).

PROCEDURE

A. Determination of an Activity Series

1. Place about 2 mL of each of the following solutions in test tubes:

$$HCl, FeSO_4, Ni(NO_3)_2, ZnSO_4$$

2. Add a clean 1 cm strip of copper to each, stir and record your observations (*1*).
3. Place about 2 mL of each of the following solutions in test tubes:

$$CuSO_4, HCl, Ni(NO_3)_2, ZnSO_4$$

4. Add a clean piece of iron to each, stir and record your observations (*2*).
5. Place about 2 mL of each of the following solutions in test tubes:

$$CuSO_4, HCl, FeSO_4, ZnSO_4$$

6. Add a clean 1 cm piece of nickel to each, stir and record your observations (*3*).
7. Place about 2 mL of each of the following solutions in test tubes:

$$CuSO_4, HCl, FeSO_4, Ni(NO_3)_2$$

8. Add a clean piece of zinc to each, stir and record your observations (*4*). Order the elements (*5*) in terms of reactivity.

B. Measurement of E° Values

9. Obtain a Carrou cell and a multimeter. You will also need one piece each of zinc, copper, nickel, iron, and graphite. These should be cleaned well using sandpaper, if necessary. Also clean the Carrou cell with soap and water and rinse well.
10. Position the Carrou cell on top of a 600 mL beaker and carry this assembly to where the solutions are dispensed. Take a 10 mL beaker as well.
11. Using a graduated cylinder, put 5 mL of each solution into the appropriately numbered cup:

Cup number	Solution
0	1 M KNO_3
1	1 M $ZnSO_4$
2	1 M $CuSO_4$
3	1 M $FeSO_4$
4	1 M $Ni(NO_3)_2$
5	1 M KI/I_2

12. Obtain about 5 mL of extra KNO_3 solution in your 10 mL beaker.
13. Back at your work area, bend each metal "electrode" into an "L" shape so that the short end dips into the solution while the longer end extends just past the edge of the cell. Place each metal into its corresponding solution. Place the piece of graphite upright in the iodide/iodine cup (number 5). DO NOT TRY TO BEND THIS!!!
14. Cut five strips of filter paper of dimensions 0.5 x 5 cm. Wet each with KNO_3 solution and place one between the central cup and each of the five outside cups so that they contact the solutions in the cups.
15. You have now set up the components for numerous galvanic cells. All that is missing is the connection between the electrodes *via* the voltmeter. This will be completed when you attach one electrical lead to the "(-)com" terminal of the multimeter and the other to the "+VWA" terminal. The meter should be set to the 3 DCV range.
16. You should now in turn measure the potential for every possible combination of cells (twelve) as indicated on the data sheet (*6*). In order to accomplish this, attach the "-ve" lead from the multimeter to the more reactive electrode (as determined from your activity series). This will be the **anode.** Then attach the "+ve" lead to the less reactive electrode, which will be the **cathode.** This should always provide you with a positive voltage (why?). Record the identity of each electrode and the voltage (*6*).
17. When you have finished your measurements, empty the solutions in your cell into the WASTE DISPOSAL BOTTLE. Wash it out well.
18. Assume that the reduction potential for your most active metal is 0.0 V. Using your observed voltages, determine the reduction potentials for the other oxidized species investigated (*7*). Determine which of the 10 species available (metals, metal cations, iodine, iodide) is the strongest oxidizing agent (*8*) and which is the strongest reducing agent (*9*).

C. Battery Construction

A Zinc Acid Battery

19. Adjust the multimeter so that it reads in the 0 – 300 mA range and connect a wire with an alligator clip on the end to each "terminal."
20. Place a graphite rod in one alligator clip, and a zinc strip in the other. These act as electrodes. Touch the two together and observe what happens (*10*).
21. Pour 75 mL of 3 M HCl into a 150 mL beaker.
22. Put the graphite rod and zinc strip into the beaker (making sure that they do not touch) and record your observations (*11*).

A Daniell Cell

23. Adjust the multimeter so that it reads volts and connect a wire with an alligator clip on the end to each "terminal."
24. Weigh a copper strip (*12*) and bend it into an "L" shape. Place the strip into a 150 mL beaker so that the "short end" lies parallel to and about 1 cm above the bottom of the beaker. Bend the "long side" of the strip over the edge of the beaker, and attach the alligator clip from the **positive** voltmeter terminal to the "long end" of the strip.
25. Weigh a zinc strip (*13*) and bend it into an "L" shape. Place the strip into the 150 mL beaker opposite the copper strip, so that the "short end" lies parallel to and about 3 cm above the copper "short end." Bend the "long side" of the strip over the edge of the beaker, and attach the "negative" alligator clip to the "long end."
26. Pour enough 1 M $ZnSO_4$ solution into the beaker so that it just covers the "short end" of the copper strip.
27. Carefully insert a long-stemmed funnel down the side of the beaker to the bottom.
28. Very carefully and slowly pour 1.0 M $CuSO_4$ (copper sulfate) solution into the funnel. If you do it correctly, it will enter underneath the zinc sulfate solution, and "push" the latter up.
29. Add enough $CuSO_4$ so that the interface between the two solutions is about half-way between the two metal strips. The final apparatus should look like the figure on the following page.
30. Leave the "battery" running for about 15 minutes, and record your observations (*14*).
31. Dismantle the "battery" and weigh the two strips (*15* and *16*).
32. Calculate the mass changes of the strips (*17* and *18*) and discuss (*19*).

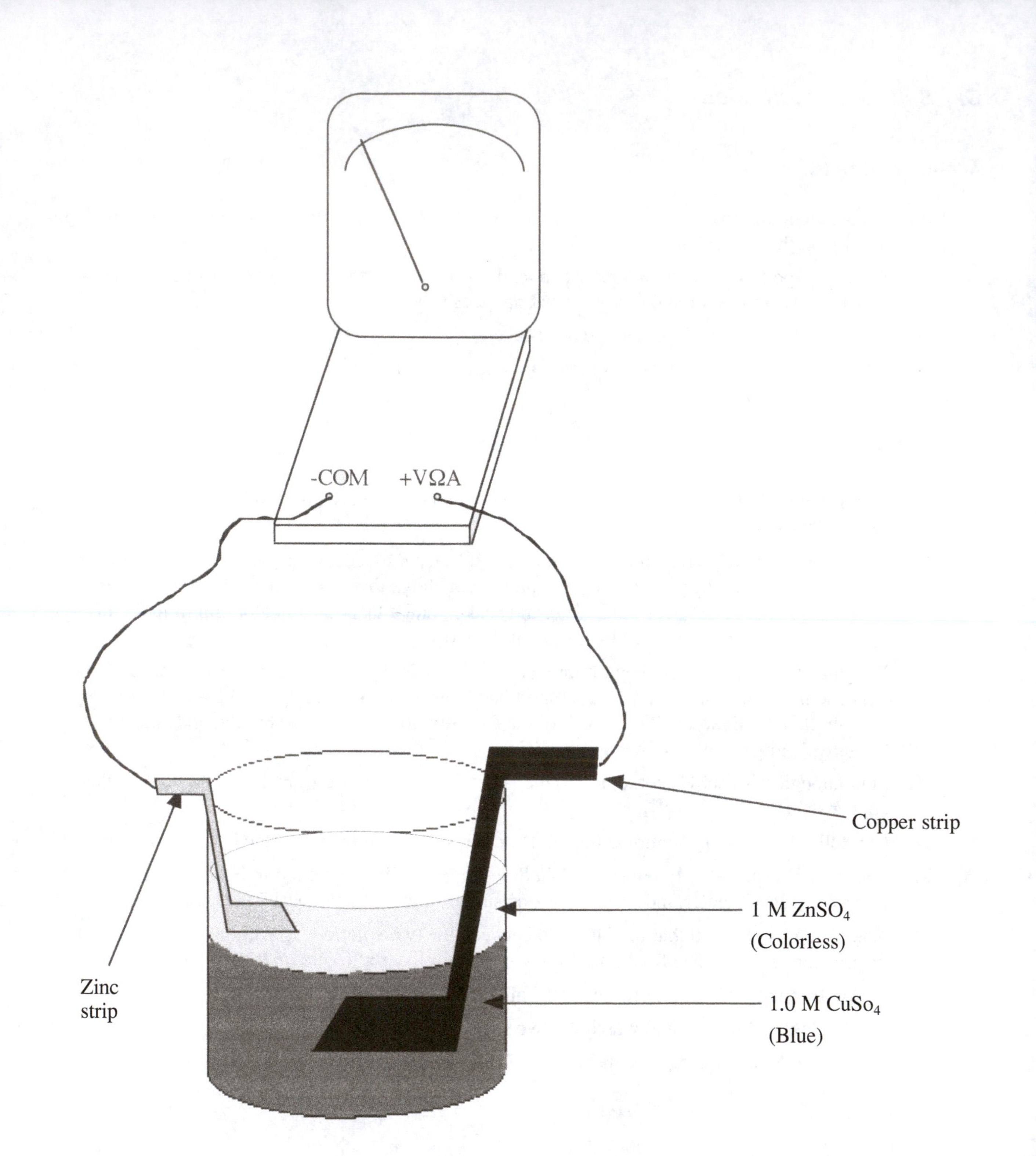
-COM
+VΩA
Copper strip
1 M ZnSO4
(Colorless)
Zinc
strip
1.0 M CuSo4
(Blue)

Electrochemistry II. Electrolysis

Introduction

Last week, you examined spontaneous electrochemical reactions and applied them in a study of galvanic cells. In today's experiment, you will carry out the reverse type of reaction, in which substances are "happy" with their electronic structure, but are forced to transfer electrons because of an applied voltage. In the first part of the experiment, you will qualitatively observe some of these electrolytic reactions (reactions that occur through electrolysis).

Another way of looking at this situation is to consider electrons as reactants or products in a chemical reaction. For example, if you take a solution of copper(II) ions and pass an electric current through the solution, you will make copper according to the equation:

$$Cu^{2+}(aq) + 2\ e^{-} \rightarrow Cu(s)$$

When studying the stoichiometry of such a reaction, it is impossible to determine the "molarity" or "molar mass" of the electrons (as is done when studying the stoichiometry of solutions or pure substances, respectively), so you need an alternative method for calculating "moles of electrons".

In order to do this, we should consider some electrical concepts. Electricity is measured in terms of charge (units of coulombs). The easiest way to calculate the quantity of electrical charge that has passed through a system is to measure the current (units of amperes or amps) passed and the time (in seconds) during which the current was passed. The relationship between these three parameters is:

$$\text{charge (coulombs)} = \text{current (amps)} \times \text{time (seconds)}$$

The charge on one electron is 1.60×10^{-19} coulombs, thus, if we know the charge, we can calculate the number of electrons used in the experiment.

Having obtained the number of electrons, it is quite simple, using the stoichiometry of the reaction, to determine the charge of one mole of electrons. In the second part of the experiment, therefore, you will carry out a reaction and calculate the charge of one mole of electrons.

Procedure

A. Some Electrolytic Reactions

1. Insert a battery into a battery-holder equipped with wires.
2. Connect an electrode (see Table) to each of the alligator clips. Be careful that the electrodes do not touch each other or you.

3. Clamp a U-tube upright, and fill it three-quarters of the way up with solution (see Table).
4. Add six drops of "Universal Indicator" to the solution and stir gently.
5. Insert the electrodes into each side of the U-tube.
6. Watch the solution carefully, looking for any changes. Record any observations in the appropriate column of the data sheet (*1*).
7. You should perform steps 1 through 6 for each of the following combinations, washing the apparatus carefully between each:

Experiment	Solution	Electrodes
1	0.1 M NaCl ✓	Graphite
2	0.1 M NaBr ✓	Graphite
3	0.1 M NaI ✓	Graphite
4	0.1 M $AgNO_3$	Graphite
5	0.1 M $CuSO_4$	Graphite
6	0.1 M $CuSO_4$	Copper

B. Determination of Faraday's Constant

8. Mark two copper strips (to distinguish them) and weigh them (*2*).
9. Attach these to the alligator clips of your apparatus, similar to that of experiment 19.
10. Half fill a 150 mL beaker with 0.01 M $CuSO_4$ solution; insert the copper strips into the solution and begin timing.
11. Read the ammeter and record the value every thirty seconds for fifteen minutes (*3*).
12. Disconnect the alligator clips, allow the electrodes to sit for a minute, then rinse them. Allow to dry on a paper towel and reweigh (*4*). (*Note.* DO NOT RUB THE ELECTRODES DRY!)
13. Perform the appropriate calculations.

Sample Calculations

2. Mass of copper strips before = 1.230 g and 1.351 g
4. Mass of copper strips after = 2.118 g and 0.463 g
5. Average current = 3 amps
6. Time = 15 minutes = 900 seconds
7. Total charge = current × time
   = 3 amps × 900 seconds = 2700 coulombs
8. Change in copper mass = (2.118 g – 1.230 g) and (1.351 g – 0.463 g)
   = 0.888 g and 0.888 g
9. Average change = 0.888 g
10. Moles of copper = mass of copper ÷ atomic mass
    = 0.888 g ÷ 63.55 g/mole = 0.0140 moles

Thus, a charge of 2700 coulombs converts 0.0140 moles of copper ions to copper metal.

From the equation, $Cu^{2+}(aq) + 2\ e^- \rightarrow Cu(s)$, we know that it takes 2 moles of electrons to produce one mole of copper:

11. Moles of electrons = moles of copper x 2
    = 0.0140 moles × 2 = 0.0280 moles of electrons.

Thus, 0.0280 moles of electrons is equivalent to 2700 coulombs of charge, and:

12. Charge of one mole of electrons = charge passed ÷ moles of electrons
    = 2700 coulombs ÷ 0.0280 moles
    = 96,430 coulombs per mole

This is the value of a FARADAY.

Synthesis of Aspirin

INTRODUCTION

In today's lab, you will make two esters of salicylic acid. Esters are molecules that contain the group shown as **(1)**. The first ester you will make is acetylsalicylic acid, formed as shown in Reaction 1. The common name for this ester is ASPIRIN. The second ester you will make is methyl salicylate—formed as shown in Reaction 2 by the reaction of salicylic acid with methanol. This ester is a component of wintergreen—a common flavoring agent.

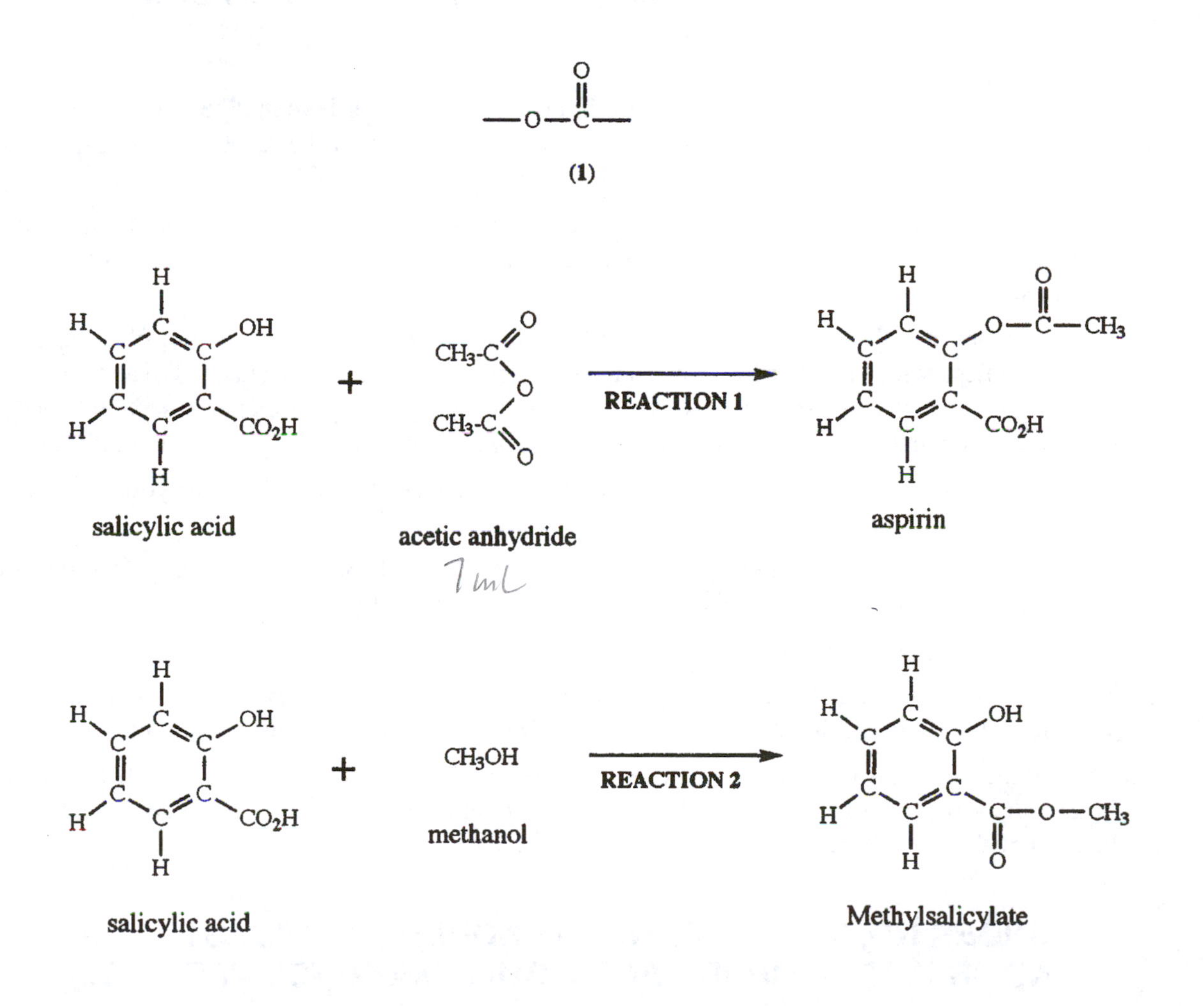

Procedure

A. *Synthesis of Aspirin*

1. Take a 125 mL Erlenmeyer flask and weigh it. Record the mass in line (*1*).
2. Add about 3 g of salicylic acid to the flask and reweigh (*2*).

CAUTION!

SALICYLIC ACID IS A SKIN IRRITANT.

3. Go to the hood and CAREFULLY add 6 mL of acetic anhydride to cover the crystals of salicylic acid.

ACETIC ANHYDRIDE IS A SKIN AND EYE IRRITANT.

4. Now add 6 drops of concentrated sulfuric acid to the mixture.

CONCENTRATED SULFURIC ACID CAUSES SEVERE BURNS. IF YOU SPLASH ANY ON YOU, WASH IT OFF AND INFORM THE TA.

5. Cover the top of the flask with aluminum foil, clamp it in a water bath and heat it for five minutes. (The surface of the water bath should be above the surface of the reactants in the flask). While the flask is heating, get about 100 mL of ice water.
6. Remove the flask from the hot bath, and add 20 mL of ice water (this will decompose any unreacted acetic anhydride). Now return to your work area and place the flask in an ice bath. You should see crystals of aspirin start to form.
7. After no more crystals form, you need to filter the solution by performing a "vacuum filtration". CLAMP a flask with a side-arm, and attach the side-arm to the rubber tube attached to an "aspirator" on the water faucet. When the water is turned on, the flow past the aspirator creates suction. Place the filter funnel into the mouth of the side-arm flask, and place a piece of filter paper inside the funnel.
8. Turn the water on, moisten the filter paper with a little water and carefully pour your reaction mixture into the funnel.
9. After you have filtered your reaction mixture, add about 10 mL of ice water to the original reaction flask, swirl around to pick up any residual solid and filter this also.
10. "Wash" the crystals with a further 10 mL of ice water.
11. This aspirin is undoubtedly very impure, and so it would need to be recrystallized before using. We will not do that in this lab. Simply leave your aspirin to air dry while doing part B.
12. When done with Part B, remove the hose from the side-arm of the flask, and turn off the faucet. Scrape the aspirin onto a preweighed piece of weighing paper (*3*). Weigh and record the mass on line (*4*) of the data sheet, and save the "aspirin" for Part C.

UNDER NO CIRCUMSTANCES SHOULD YOU INGEST YOUR ASPIRIN. IT IS IMPURE AND WOULD MAKE YOU VERY ILL.

B. *Synthesis of Methyl Salicylate*

13. Place 1 g of salicylic acid and 5 mL of methanol (methyl alcohol) in a test tube.

CAUTION!

SALICYLIC ACID IS A SKIN IRRITANT.

14. Carefully add 3 drops of concentrated sulfuric acid and then heat in a water bath (at a temperature of about 70 °C) for about 15 minutes. Note the odor (IN THE CORRECT MANNER). Record observations (*7*).

CONCENTRATED SULFURIC ACID CAUSES SEVERE SKIN BURNS. INFORM THE TA OF ANY SPILLS.

C. *Analysis of Aspirin*

15. Take three small test tubes and add 1 mL of ethanol and one drop of $FeCl_3$ solution to each.
16. To the first tube, add a pinch of salicylic acid using a spatula.
17. To the second, add a pinch of your aspirin using a spatula.
18. Shake each tube and record your observations (*8*).
19. If both tubes react in the same way, it means your aspirin is very unpure. There is some powdered “real” aspirin laid out to use in step 17 if you need it.

Separation by Paper Chromatography

INTRODUCTION

Powerful tools and techniques are available for the identification of chemical substances. For most of these, however, it is necessary to start with a pure substance in order to perform a proper analysis. Thus, techniques for separation are almost as important as those used for synthesis or analysis. Among these, probably the most familiar are distillation (which takes advantage of differences in vapor pressures) and recrystallization (which takes advantage of differences in solubilities). Another method, the one to be studied in this experiment, is known as **chromatography**.

In chromatography, a mixture is usually dissolved in a **moving** (or **mobile**) phase, which moves in contact with a **stationary** phase. As different substances in the mixture should have different solubilities (or affinities) in the two phases, their progress will proceed at different rates. For example, consider the example shown in Figure 22–1a. A spot of a mixture which consists of three components, A, B, and C put on a piece of paper (which will be the stationary phase). This paper is then placed vertically in a beaker of water (which will act as the mobile phase). If all the components are soluble in water, the spot will "dissolve" and the "solution" will begin to move up the paper via capillary action. However, A, B, and C will each have different solubilities in water and affinities for paper. In this action, we assume that the affinity for paper decreases in the order $A > B > C$, and the solubility in water decreases so that $C > B > A$. Thus, C will move more quickly than B, which will move more quickly than A. So, after a period of time, the experiment will look like that shown in Figure 22–1b as the components become separated from each other (you should note that the separation will increase as more time passes, Figure 22–1c).

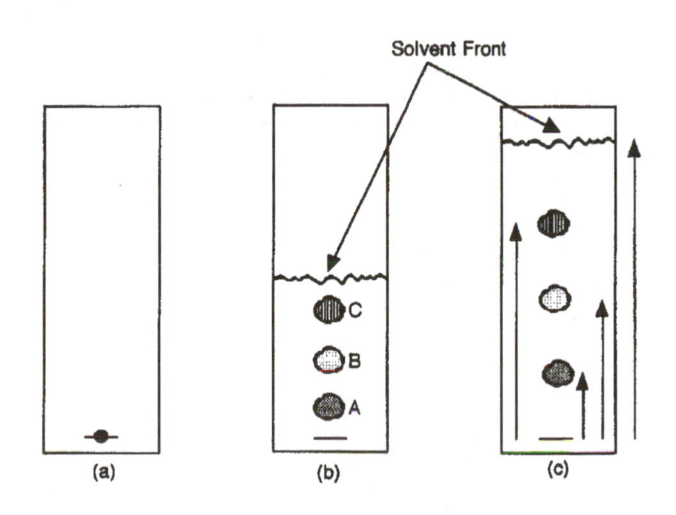

Figure 22–1

In order to express quantitatively the differences in separation, the retention factor (or R_f) is used. This is a fractional number that depends on solute and solvent as well as temperature:

$$R_f = \text{distance moved by solute} / \text{distance moved by solvent}$$

These distances are drawn for each substance in Figure 22–1c. If one knows R_f values for a number of solutes in a given solvent, a particular solute may be identified by comparing R_f values.

A final experimental issue concerns the methods used to actually see the spots corresponding to the different components of the mixture. In the original application of chromatography (developed by Tswett in 1903), different plant pigments were separated according to their color (hence the name which comes from the word chroma or color). However, most substances are colorless. In such cases, different agents may be added to the chromatogram in order to "colorize" the sample.

In today's experiment, you will first determine the R_f values for a mixture of acid/base indicators. These are colored in both acidic and basic solution. In the first part of the lab, you will separate the indicators using a basic aqueous solution as the mobile phase. You will then expose the separated spot to acid vapors and examine the changing of their colors. Finally, you will use your known R_f values to analyze an unknown mixture.

Procedure

1. Label four small test tubes for the different indicators: MO (methyl orange), MR (methyl red), MV (methyl violet) and PR (phenol red). Obtain about 1 mL of each indictor in the appropriate test tube. Also collect a sample of your unknown mixture (your TA will assign you a number).
2. In a fifth test tube, mix small equal amounts of the four indicators.
3. Obtain a "developing chamber" (your 400 mL beaker) and add enough "developing solution" (the mobile phase, 0.1 M NaOH) to give a depth of about 5 mm. Also collect a piece of clear plastic wrap large enough to cover the top of the beaker.
4. Obtain a piece of filter paper that should have been cut to the correct dimensions (approximately 10×20 cm).
5. Align the filter paper so that the long edge is horizontal. Draw a pencil line lightly about 1 cm above the bottom and mark 6 evenly spaced "dots" along this line. Label these as shown below, for the four indicators, a known mixture (KN) and an unknown mixture (UNK).

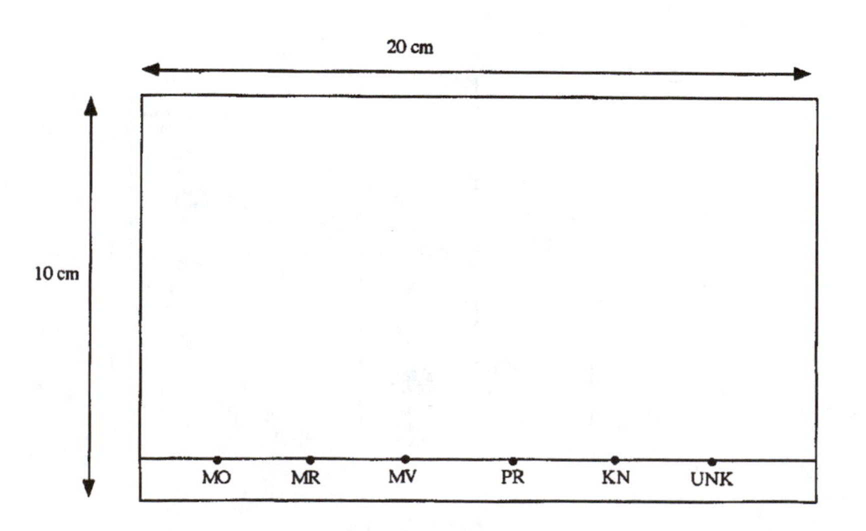

6. Draw up a very small drop of the first indictor into a clean capillary and touch the tip to the appropriate dot on the filter paper. THIS DOT SHOULD BE NO BIGGER THAN 2 mm IN DIAMETER. You might find it useful to practice your technique a few times on a scrap of filter paper or a paper towel before "doing it for real."
7. Repeat step 6 for the remaining three known indictors, the known mixture and the unknown mixture. YOU SHOULD CLEAN THE CAPILLARY WELL BETWEEN EACH SAMPLE.
8. Form the paper into a cylinder so that the two ends are almost (but not quite) touching and staple the ends together.
9. Place the cylinder into the developing chamber with the sample line at the bottom (BE SURE THAT THE SURFACE OF THE DEVELOPING SOLUTION IS LOWER THAN THE PENCIL LINE).
10. Cover the developing chamber with the plastic wrap. Leave beaker undisturbed until the mobile phase has moved to within about 1 cm of the top (about 30 minutes).
11. Remove the cylinder, measure the distance moved by the solvent (mobile phase) and record (*1*). Mark the center of each spot using a pencil. For each indictor spot, record the color (*2a*) and measure the distance each spot has moved from the starting line and record (*2b*). Calculate the R_f value (*2c*) by dividing the numbers recorded in *2b* by the number recorded in line *1*.
12. For each mixture [the known (*3*) and the unknown (*4*)] measure and record on the data sheet the distances moved by each spot in order of increasing distance moved (*3a* and *4a*) and assign to each spot the identity of the appropriate indicator (*3b* and *4b*).

CAUTION!

IN THE HOOD

13. Place a few drops of concentrated HCl in a second beaker and hold your chromatogram above the liquid in the fumes for a few seconds.

CONCENTRATED HCL IS EXTREMELY CORROSIVE. THE VAPORS ARE EXTREMELY IRRITATING TO YOUR EYES. WASH ANY CONTAMINATED AREAS OF SKIN VERY WELL AND INFORM YOUR TA.

14. Record the new colors of each indicator spot (*2d*).

Name ______________________________ TA ______________________________

DATA SHEET—EXPERIMENT 22

Separation by Paper Chromatography

1. Distance traveled by solvent ______

2. *Known indictors*:

	MO	MR	MC	PR
a. color	______	______	______	______
b. distance moved	______	______	______	______
c. R_f value (*2b / 1*)	______	______	______	______
d. color in HCl	______	______	______	______

3. *Known Mixture*:

	Spot 1	Spot 2	Spot 3	Spot 4
a. distance moved	______	______	______	______
b. identity	______	______	______	______

4. *Unknown Mixture*:

	Spot 1	Spot 2	Spot 3	Spot 4
a. distance moved	______	______	______	______
b. identity	______	______	______	______

POSTLAB

1. Suppose that you attempted to separate two indicators using paper chromatography using 0.1 M NaOH but were unsuccessful. Explain what you could do to improve the separation:

2. A particular compound has an R_f value of 0.95. Is it more strongly held on the paper or in the solution? Explain.

3. Explain why we chose acid-base indicators as the substances in this experiment instead of more common substances such as sugar or salt, *etc.*

4. Explain why it is important not to permit the solvent front to reach the top of the chromatogram.

5. Why is it important to keep the indicator spots very small?

Redox Reactions

INTRODUCTION

A very important class of chemical reactions are those that involve the transfer of electrons. In such reactions, one species loses electrons (is **oxidized**), the other gains the electrons (is **reduced**). As any such reaction must involve both forms of activity, they are called **reduction-oxidation** or **redox** reactions.

When dealing with species that have gained or lost electrons, it helps to have a system whereby one can keep track of the electron movement. One such system is to assign **oxidation numbers** to atoms. The oxidation number can be thought of as the "charge" on the end species. Thus, if an atom loses an electron, it will have a charge of +1; its oxidation number, therefore, is +1. Similarly, if an atom gains an electron, it will have a charge and an oxidation number of –1. For example, sodium will almost always lose one electron to become Na^+ so the sodium atom has an oxidation number of +1. Oxygen will almost always gain two electrons to become O^{2-}, thus oxygen usually has an oxidation number of –2. One can understand redox equations by assigning oxidation numbers to all the atoms. As electrons move between species, the "charge" and, thus, oxidation numbers of those species will change (the oxidized atom will increase in oxidation number, while the reduced species will decrease in oxidation number). Accordingly, one can understand the redox reaction by following the change in oxidation number.

Many atoms have "preferred" oxidation numbers, which usually relate to the number of electrons the atom must lose or gain to attain the electron configuration of an ideal gas. This makes it rather easy to assign oxidation numbers to such atoms. The transition metals, on the other hand, can be "happy" in several oxidation states. As an illustration of the principles involved, in this experiment you will examine the oxidation numbers and redox behavior of one such transition metal—manganese.

In the first part of the lab, you will generate manganese in four different oxidation states. Each of these oxidation states has a distinctive color in aqueous solutions. In the second part of the lab, you will investigate an oxidation/reduction reaction using titration techniques. There are two titrations to be performed—one where you "standardize" (*i. e.*, determine the exact concentration) of a potassium permanganate solution and the second, in which you use the concentration of the permanganate to determine the concentration of an unknown solution of oxalate ions.

Permanganate ions (MnO_4^-) are superb oxidizing agents as they contain manganese in a very high oxidation state (+7). When added to an acidic solution containing oxalate ions ($C_2O_4^{2-}$), they will oxidize the oxalate to carbon dioxide according to the following equation:

$$5\ C_2O_4^{2-}(aq) + 2\ MnO_4^-(aq) + 16\ H^+(aq) \rightarrow 10\ CO_2(g) + 8\ H_2O(l) + 2\ Mn^{2+}(aq)$$

This reaction may be followed through the color change of the ions using results from Part A.

$NaHSO_3 \rightarrow Na_2S_2O_3$

PROCEDURE

A. *Oxidation States of Manganese*

Post Lab Q#3

1. Take three 100 mL beakers and add 10 mL of 0.005 M $KMnO_4$ solution to each. This solution contains manganese with a +7 oxidation number. Record the color (*1*).
2. To one beaker, add 8 mL of 1 M NaOH solution and stir.
3. While stirring, add slowly 15 mL of 0.01 M $NaHSO_3$ solution. The final color is due to MnO_4^{2-}, which contains manganese with a +6 oxidation number. Record the color (*2*).
4. To the second beaker from step 1, add slowly 15 mL of 0.01 M $NaHSO_3$ while stirring. The solid that forms is MnO_2 (manganese with a +4 oxidation number). Record the color (*3*).
5. To the third beaker, add 10 mL of 1 M H_2SO_4 and stir.
6. While stirring, add slowly 25 mL of 0.01 M $NaHSO_3$ solution. The final color is due to Mn^{2+}—manganese with a +2 oxidation number. Record the color (*4*).

B. *Standardization of $KMnO_4$ Solution*

7. Weigh a 250 mL Erlenmeyer flask (*5*).
8. Add between 0.2 and 0.3 g of sodium oxalate. Reweigh and record the exact mass (*6*).

CAUTION!

SODIUM OXALATE IS A POISON—DO NOT INGEST!

9. Add 100 mL of water and 20 mL 3 M sulfuric acid. Swirl until all of the solid has disappeared.

SULFURIC ACID IS A STRONG ACID AND IS CORROSIVE.
SEE YOUR TA IF YOU SPILL ANY!

10. Fill a buret with the $KMnO_4$ solution of an unknown concentration. Add about 15 mL of this to your sodium oxalate solution, swirl and then allow to stand until the color has disappeared. If you have problems, repeat steps 7 through 10 using less $KMnO_4$.

POTASSIUM PERMANGANATE WILL STAIN SKIN ON CONTACT.

11. Heat the solution gently to about 80 °C. At this point, remove the heat and resume titration. You should continue until a pale pink color (indicative of unreacted permanganate) persists for at least 30 seconds.
12. Record the TOTAL volume of $KMnO_4$ added (*7*).
13. Repeat the entire procedure.
14. Now perform the appropriate calculations to find the $KMnO_4$ concentration.

C. Determination of the Unknown Oxalate Solution

15. Your TA will give you an unknown oxalate solution. Pipet exactly 25.0 mL (*15*) of the unknown into a clean 250 mL Erlenmeyer flask and add 75 mL of water and 20 mL of 3 M H_2SO_4. Mix these by swirling and repeat the titration procedure as described in Part B (from step 10). Record the total volume of $KMnO_4$ solution added (*16*).
16. Repeat the process and then perform the appropriate calculations to find the oxalate concentration.
17. Obtain the class data (*21*) and calculate the mean (*22*) and standard deviation (*23*).

Check table of reagents

SAMPLE CALCULATIONS

B. Standardization of $KMnO_4$

When I did the experiment, my data were:

5. Mass of flask (g) 68.000
6. Mass of oxalate and flask (g) 68.250
7. Volume of permanganate added (mL) 30.0
8. Mass of oxalate = (mass of oxalate and flask) – (mass of flask)
   = 68.250 g – 68.000 g = 0.250 g
10. Moles of oxalate = mass of oxalate ÷ molar mass
   = 0.250 g ÷ 134.00 g/mole = 1.87×10^{-3} moles
11. Equiv. moles of MnO_4^- :
   From the stoichiometric equation, 5 moles of oxalate react with 2 moles of permanganate.
   moles of oxalate = moles of permanganate x 5/2
   moles of permanganate = $2/5 \times$ moles of oxalate = 7.48×10^{-4} moles
12. Liters of MnO_4^- solution = mL of MnO_4^- ÷ 1000 = 0.030 L
13. Molarity of MnO_4^- solution = moles ÷ liters
   = 7.48×10^{-4} moles ÷ 0.030 liters = **0.025 M**

C. Determination of the Unknown Oxalate Solution

The calculations for this are essentially the same. You know the permanganate concentration, and must calculate that of the oxalate. Use your intelligence and basic chemical knowledge!!!!

Acid-Base Titrations: The Analysis of Vinegar

Introduction

Titration is the process in which one measures the volume of a solution of known concentration (the **titrant** solution) which is required to react completely with a given volume of an unknown solution (the **analyte** solution). The titrant is usually contained in a buret and is added slowly to a measured volume of analyte until exactly the right amount of titrant has been supplied to complete the reaction (the stoichiometric point or **equivalence point**). This point is sometimes marked by a color change in the solution due to the complete disappearance of a colored reactant. Alternatively, it is often detected by indictors, which are substances added to the reaction that change color in the appropriate chemical environment.

In today's experiment, you will perform several titrations in which the titrant is a base (NaOH) and the analyte solutions are acids. The general reaction that will occur is, therefore, a neutralization, the net ionic equation for which is:

$$H^+ + OH^- \rightarrow H_2O$$

At the start, the analyte solution contains H^+ ions (which are acidic). As OH^- ions are added, these neutralize the acid. Thus, before the equivalence point, the analyte solution is acidic; after the equivalence point, any further titrant added is basic. We shall use acid-base indictors to show the equivalence point. These compounds are complex organic molecules which have one color in acidic solution and a different color in basic solution. Thus, when an indicator is added to an acid-base titration, a change in color (the end-point) is used to mark the equivalence point. Different indicators change color at different concentrations of H^+, so it is important to pick an indicator whose end-point corresponds with the equivalence point of the appropriate titration.

In today's experiment, you will first examine qualitatively several different indicators. You will then perform two different titrations to determine the concentrations of an unknown HCl solution and an unknown acetic acid (CH_3COOH) solution, otherwise known as vinegar.

Procedure

1. Obtain about 50 mL of NaOH and 25 mL of the HCl solution assigned by your TA.

A. *Acid-base Indicators*

2. Add about 2 mL of deionized water into each of three small test tubes.
3. Add one drop of phenolphthalein to one tube, one drop of methyl orange to second, and one drop of bromcresol purple to the third. Record the colors (*1a*).

4. Add 2 drops of NaOH solution to each test tube and record the color (*1b*).
5. Add HCl dropwise to each tube. Count and record (*1c*) the number of drops required to produce a color change. Record the new color in acidic solution (*1d*).
6. Which indicator has the most acidic end-point (*2*); which has the most basic end-point (*3*).

B. Titration of an Unknown HCl solution

7. Record the number of your unknown solution (*4*).
8. Obtain a ring stand, a buret clamp and two burets.
9. Rinse one buret with your HCl solution, then clamp and fill it with what remains of the sample you collected in step 1.
10. Rinse the second buret with NaOH solution, then clamp and fill it with what remains of the sample you collected in step 1. Record the exact concentration of the NaOH (*5*).
11. Add about 10 mL of the HCl solution into a 125 mL Erlenmeyer flask. Record the exact volume (*6*). Add about 25 mL of deionized water and 2–3 drops of phenolphthalein. Place the flask under the NaOH buret.
12. Record the volume of NaOH in the buret (*7*).
13. Slowly add NaOH into the flask, swirling as you do this. You might want to use a magnetic stirrer (depending on your TA). You should see a pink color forming that disappears as you swirl the solution. The longer that it takes for this color to disappear, the closer you are to the equivalence point.
14. Continue to add NaOH until the pink color does not disappear. Record the final buret reading (*8*).
15. Calculate the concentration of the HCl solution (*10*).
16. Repeat steps 11 through 15, refilling the NaOH buret if necessary (remember to include this in your calculations!).
17. Calculate the average value of the HCl concentration (*11*).

C. Titration of Acetic Acid in a Vinegar Solution

18. Obtain about 25 mL of vinegar solution.
19. Wash out and rinse the "HCl buret" with the vinegar solution, then clamp and fill it with what remains of the sample you collected in step 18.
20. Refill the NaOH buret if necessary.
21. Repeat steps 11 through 17 above, replacing HCl by vinegar throughout. Record your values in lines *12* through *17* on the data sheet, as appropriate.
22. Give your average acetic acid concentration to your TA. After lab, record the class data (*18*) and calculate the mean (*19*) and standard deviation (*20*).

Sample Calculations

You should do your calculations using the facts that:

a) molarity = moles of solute / liters of solution

b) in both reactions, 1 mole of NaOH reacts with 1 mole of acid.

When I did the HCl titration, my data were:

5.	NaOH concentration:	0.252 M
6.	Volume of HCl solution	10.0 mL
7.	Initial Buret reading	0.0 mL
8.	Final Buret reading.	15.2 mL
9.	Volume of NaOH used (*8 – 7*)	15.2 mL

We want the molarity or moles/liter of the HCl solution. We already know the number of liters (line 6, = 0.0100 L), so we need moles of HCl.

To get moles of HCl, we need to know moles of NaOH (as the reaction is 1:1).

To get moles of NaOH, we use the molarity (line *5*) and volume (from lines *7* and *8*).

molarity of NaOH = (moles of NaOH) / (liters of NaOH solution)

0.252 M = (moles of NaOH) / (0.0152 L)

moles of NaOH = $0.252 \times 0.0152 = 0.00383$ moles

moles of HCl = 0.00383 moles

molarity of HCl = (moles of HCl) / (liters of HCl)

molarity of HCl = (0.00383) / (0.010) = **0.383 M**

Name ______________________ TA ______________________

DATA SHEET—EXPERIMENT 24

Acid-Base Titrations: The Analysis of Vinegar

A. Acid-Base Indicators

1.		phenolphthalein	methyl orange	bromcresol purple
a)	Color in water	______	______	______
b)	Color in basic solution	______	______	______
c)	Number of drops of acid	______	______	______
d)	Color in acidic solution	______	______	______
2.	Indicator with most acidic end-point?	______		
3.	Indicator with most basic end-point?	______		

B. Titration of HCl Solution

4.	Unknown Number:	____	
5.	NaOH concentration (M)	____	____
6.	Volume of HCl solution (mL)	____	____
7.	Initial buret reading (mL)	____	____
8.	Final buret reading (mL)	____	____
9.	Volume of NaOH used (mL) (*8 – 7*)	____	____
10.	Concentration of HCl solution (M)	____	____
11.	Average HCl concentration (M)	____	

C. Titration of Acetic Acid Solution

12.	Volume of acetic acid solution (mL)	____	____
13.	Initial NaOH buret reading (mL)	____	____
14.	Final NaOH buret reading (mL)	____	____
15.	Volume of NaOH used (*14 – 13*)	____	____
16.	Concentration of vinegar solution solution (M)	____	____

17. Average vinegar concentration (M) _____

18. Class Data

_____	_____	_____	_____	_____	_____
_____	_____	_____	_____	_____	_____
_____	_____	_____	_____	_____	_____
_____	_____	_____	_____	_____	_____
_____	_____	_____	_____	_____	_____

19. Mean acetic acid concentration (M) _____

20. Standard deviation of acetic acid concentration. _____

POSTLAB

1. What is a meniscus? Show with a sketch. What part does the meniscus play in this experiment?

2. Why is it not necessary to start a titration with a volume in the buret that reads exactly 0.0 mL?

3. What is the purpose of the phenolphthalein used in parts B and C of this experiment? Explain briefly how it works.

4. Compare the class data with yours. What is your percent error? Where might this have arisen?

5. Using the class value for the concentration of acetic acid in the vinegar solution, calculate the mass percent of acetic acid in common vinegar. Compare this value to that of a vinegar value in your favorite supermarket.

Antacid Titration

INTRODUCTION

You have become familiar with the technique of titrations—in which the amount of one reactant needed to react with another is determined. In today's experiment, you will measure the efficiency of two "antacids" by titrating them with acid to the "end" of the reaction.

An "antacid" is simply a base that neutralizes excess stomach acid. The usual form of an antacid is one that produces bicarbonate ion (HCO_3^-). This ion reacts with acid to produce water and carbon dioxide, the second of which is often released by the body orally! Thus, the most effective antacid would be simple baking soda (sodium bicarbonate) or an equivalent. However, baking soda does not have a pleasing taste, so commercial antacids also contain extra, inactive ingredients. In this lab, you will compare two common antacids—Rolaids and Tums—to decide which of the two is more effective per gram (*i. e.*, neutralizes the more acid per gram).

To do this, you will essentially measure the amount of acid needed to "neutralize" the antacid. Unfortunately, this is not as simple as it sounds. Bicarbonate is "amphiprotic" (*i. e.*, can act as an acid or a base), so it forms a buffer system in water. As buffers are resistant to change in pH, if you were simply to titrate with HCl, your results would be in error. Accordingly, you will mix the antacid with a large excess of acid, to overwhelm the buffer, and then titrate this acidified mixture with sodium hydroxide to measure how much acid was not neutralized by the antacid. This technique is called a back titration.

PROCEDURE

1. Weigh a 250 mL Erlenmeyer flask (*1*).
2. Add about 0.7 g of powdered antacid to the flask and reweigh (*2*).
3. Add 50.0 mL of 0.1 M HCl from a buret into the flask. You should record the EXACT molarity of the HCl on the data sheet (*3*) and the volume added (*4*). Swirl the mixture.
4. Heat the solution to a boil and maintain the heat for about a minute. This ensures both that the antacid is dissolved and that any dissolved CO_2 in the water is removed.
5. Add 5 drops of bromphenol blue. This compound is an indicator (weak acid) that is blue in a basic solution and yellow in an acid. Such a compound, whose color depends on the pH, is called an acid-base indicator.
6. If the solution is blue, it means that there was too much antacid. Add a further 25.0 mL of HCl and repeat step 4. Do not forget to add the extra volume to line (*4*) on the data sheet.
7. Rinse a clean buret with 0.1 M NaOH solution. Record the exact molarity of the NaOH on the data sheet (*5*).

8. Fill the buret with NaOH solution and record the exact volume (*6*).
9. Titrate the antacid sample with the NaOH. Your end point will be when the color changes to and remains blue for at least 20 seconds. Record the final volume left in the buret (*7*).
10. Carry out the experiment twice for each antacid and perform the appropriate calculations.

Sample Calculations

When I did the experiment, I obtained the following data:

1. Mass of flask (g) = 77.357 g
2. Mass of flask + antacid (g) = 78.041 g
3. Concentration of HCl (M) = 0.0997 M
4. Volume of HCl (mL) = 50.0 mL
5. Concentration of NaOH (M) = 0.1040 M
6. Initial volume of NaOH (mL) = 0.0 mL
7. Final volume of NaOH (mL) = 24.0 mL
8. To determine the moles of HCl added

Molarity	= moles ÷ liters
0.0997	= moles ÷ 0.0500 L
moles (HCl)	= 0.0997×0.0500
	= 4.985×10^{-3} moles

9. To determine the moles of NaOH added

Molarity	= moles ÷ liters
0.1040	= moles ÷ 0.0240 L
moles (NaOH)	= 0.1040×0.0240
	= 2.496×10^{-3} moles

10. To determine the active moles of antacid

Remember what we did. We took a certain amount of antacid, and added a huge excess of acid. We then used up the rest of the acid by titration with NaOH. Thus, the total acid added is the same as the total of the two bases (NaOH and the antacid). *i. e.*,

moles of acid	= active moles of antacid + moles of base
$4.985 \text{ x } 10^{-3}$	= active moles of antacid + 2.496×10^{-3}
active moles of antacid	= $4.985 \times 10^{-3} - 2.496 \times 10^{-3}$
	= 2.489×10^{-3} moles
11. Grams of antacid (g)	= (mass of flask + antacid) – (mass of flask)
	= 78.041 g – 77.357 g
	= 0.684 g
12. Active moles of antacid per gram	= moles of antacid ÷ grams
	= 2.489×10^{-3} moles ÷ 0.684 g
	= 3.64×10^{-3} moles / gram

Thus, one gram of this antacid will neutralize 3.64×10^{-3} moles of acid.